0342165 −9

U.
mo
for
Boo
ano

DUE F

DUE FOR

8 JUL

DUE

-7

Springer Advanced Texts in Life Sciences

David E. Reichle, Editor

Leo J. Fritschen
Lloyd W. Gay

Environmental Instrumentation

With 66 Figures

Springer-Verlag
New York   Heidelberg   Berlin

Leo J. Fritschen
College of Forest Resources
University of Washington
Seattle, Washington 98195
USA

Lloyd W. Gay
School of Renewable Natural Resources
University of Arizona
Tucson, Arizona 85721
USA

Series Editor:
David E. Reichle
Environmental Sciences Division
Oak Ridge National Laboratory
Oak Ridge, Tennessee 37830
USA

Library of Congress Cataloging in Publication Data

Fritschen, Leo.
  Environmental instrumentation.

  (Springer advanced texts in life sciences)
  Bibliography: p.
  Includes index.
  1. Environmental monitoring—Instruments. 2. Physi-
cal instruments. I. Gay, Lloyd Wesley, 1933–
II. Title.
TD170.2.F73   550′.28   79-10437

ISBN 0-387-90411-5 Springer-Verlag   New York
ISBN 3-540-90411-5 Springer-Verlag   Berlin   Heidelberg

Preface

The rapid increase in environmental measurements during the past few decades is associated with (1) increasing awareness of the complex relations linking biological responses to atmospheric variables, (2) development of improved data acquisition and handling equipment, (3) the application of modeling to environmental problems, and (4) the implementation of large, cooperative studies of international scope.

The consequences of man's possible alteration of the environment have increased our interest in the complex nature of biological responses to meteorological variables. This has generated activity in both measurements and in the application of modeling techniques. The virtual explosion of modeling activity is also associated with the development of large computers. The testing of these models has demonstrated the need for more, different, and better environmental data. In addition, technological developments, such as integrated circuits, have reduced the cost, power consumption, and complexity of data acquisition systems, thus promoting more environmental measurements.

The emergence of scientific cooperation on a global scale has increased measurement activities markedly. The International Geophysical Year (1958) has been followed by the International Hydrologic Decade, the International Biological Program, the Global Atmospheric Research Program, and a host of environmental studies of a regional nature that have all emphasized field data collection.

With few exceptions, space-age technology has led to improved methods for data recording and handling, rather than changes in instruments used to sense the environment. Thus, while recording methods have progressed from mechanically driven pens to data systems coupled with on-line computers, the same basic sensors have remained in use.

These developments have made it easier to collect large quantities of data, but all too frequently sensors are not properly exposed, electrically isolated, or even compatible with the recording instruments. Vast quantities of recorded data have often turned out to be invalid.

Courses on environmental instrumentation are not common on university campuses, despite the need for training on this topic. Earlier books on the subject, such as *Meteorological Instruments* by Middleton and Spilhaus, are out of date and out of print. This book is designed to be used as a text for advanced students and a guide or manual for researchers in the field. Our purpose is to present the basic theory of environmental variables and transducers, report our experiences on methodology and use, and provide certain essential tables. The user is expected to have a basic physics and mathematics background and to be knowledgeable in the area of his speciality.

We will concentrate on the principles that govern the use of sensors and the operation of recorder systems as these are less rapidly affected by technological process. The applications will use currently available equipment.

September, 1979   Leo J. Fritschen
                                                     Lloyd W. Gay

Contents

List of Symbols

Symbol	Unit	Definition
a		constant, low temperature, absorption coefficient, ratio
A		constant, aspect ratio, intercept, variable metal, analog domain
A	$°C^{-1}$	thermodynamic psychrometric constant
A	m^2	area
A_c	m^2	convective area
A_r	m^2	radiational area
A_0		Ferrel psychrometric constant
b		constant, midrange temperature
B		amplitude, constant, slope, variable, metal
B	$W(m^2\ sr)^{-1}$	steradiancy
c	$J(kg\ K)^{-1}$	specific heat
c		speed of light, constant
c	$m\ s^{-1}$	speed of sound in still air
c		high temperature
c_p	$J(kg\ K)^{-1}$	specific heat at constant pressure
C		constant, slope, variable, metal, correction factor
C	$J(m^3\ K)^{-1}$	heat capacity
d	m	diameter, delay distance
D	$W\ m^{-2}$	diffuse radiation flux density

D	V	thermoelectric potential
D	m	distance constant
D		variable, digital domain
e		natural logarithm
e	Pa	vapor pressure
E	V	applied voltage, thermal emf
E		edge correction, combined value, error
E	W m^{-2}	irradiance
E_i		Einstein (mole of photons)
E_i	lm m^{-2}	illumance
E_v		velocity error
f	Hz	sound wave frequency
F	kg(ms^2)$^{-1}$	force
F		variable
F	lm	luminous flux
F	°F	Fahrenheit degree
g	m s^{-2}	acceleration due to gravity
g	V m^2 W^{-1}	calibration coefficient
Gr		Grashof number
G	W m^{-2}	heat flow through a medium
G		variable
h	m	height
h	W(m^2 K)$^{-1}$	convective heat transfer coefficient
h		Plank's constant
h		damping ratio
H	W m^{-2}	sensible heat flux density
H		constant
i	A	current, electrical
I	kg^2 m^{-1}	moment of inertia
I	W sr^{-1}	radiant intensity
I	W m^{-2}	direct-beam solar radiation perpendicular to sun's rays
I_i	lm sr^{-1}	luminous intensity
J		mechanical equivalent of thermal energy
J		constant
k		Boltzmann's constant, coefficient
k		wavelength dependent coefficient
k	m^2 s^{-1}	thermal diffusivity

K	W m^{-2}	solar radiation flux density, constant
K_m		constant
K_r		view factor
K_v		vane quality factor
$K\uparrow$	W m^{-2}	reflected solar radiation
$K\downarrow$	W m^{-2}	global solar radiation
$K*$	W m^{-2}	net solar radiation
K	K	Kelvin degree
l	m	length
L	J kg^{-1}	latent heat of vaporization
L		lead
L	W(m^2 sr)$^{-1}$	radiance (radiant intensity per unit area)
$L\downarrow$	W m^{-2}	longwave atmospheric radiation
$L\uparrow$	W m^{-2}	longwave terrestrial radiation
$L*$	W m^{-2}	net longwave radiation
m	kg	mass
M		aerodynamic damping, measured value
M	kg	molecular weight
M	W m^{-2}	radiant emittance
M_i	cd sr m^{-2}	luminous emittance
n		number, number of moles, eddy shedding frequency
n	m^{-1}	wave number
Nu		Nusselt number
N	V °C^{-1}	thermoelectric power
o		vertex
P	W	electrical power
P	Pa	pressure
Pr		Prandtl number
P		coil, physical domain, potential
q	kg kg^{-1}	specific humidity
Q		quantum of radiation, entity, quantity of heat
Q	W m^{-2}	all wave radiation
$Q*$	W m^{-2}	net radiation flux density
r		recovery factor
r	kg kg^{-1}	mixing ratio
r	m	radius
r_w	m	radius of counter weight

R		universal gas constant
R	Ω	electrical resistance
Re		Reynolds number
s		extinction coefficient
s	m	vertical span of air foil
S	$W\ m^{-2}$	direct-beam radiation
S		Strouhal number, cubical expansion coefficient, coil
S	m^2	area of air foil
t	s or min	time
T	°C or K	temperature
T	$kg\ m^2\ s^{-2}$	torque
T_a	°C	air temperature
T_d	°C	dew-point temperature
T_s	°C	surface temperature
T_w	°C	wet-bulb temperature, wall
u		unknown
U	$m\ s^{-1}$	wind speed
U		relative humidity
v		true value
V	m^3	volume
V_λ		relative luminous efficiency
$\overline{X}$		mean of sample population,
X		Wien's constant, volume fraction
$\overline{Y}$		mean of infinite population
y		variable
z	m	depth or height
α		first order coefficient, absorption coefficient, attenuation factor
β		second order coefficient, thermistor constant, thermal expansion coefficient
γ		reflection coefficient
γ	$Pa\ °C^{-1}$	psychrometric constant

Δ	Pa $°C^{-1}$	slope of saturation vapor pressure curve
Δf		Doppler shift
ε		emissivity, ratio of mole weight of water vapor to dry air (0.622), ratio of transducer conductivity to medium conductivity
θ		angle
θ	$°K$	temperature
λ	s	time constant
λ	$W(m\ K)^{-1}$	thermal conductivity
λ_d	m	damped wavelength
λ_n	m	natural wavelength
λ	μm	wavelength
μ		dynamic viscosity
Ω		angle
ω	sr	solid angle
ω	s^{-1}	angular frequency
ν	Hz	frequency
ν		kinematic viscosity
ν	$m^3\ kg^{-1}$	specific volume
ρ	$kg\ m^{-3}$	density
ρ_v	$kg\ m^{-3}$	absolute humidity
σ	$W\ m^{-2}\ K^{-4}$	Stefan–Boltzmann constant
σ	$g\ s^{-1}$	surface tension
τ		transmission coefficient, time constant
Φ	W	radiant flux
ϕ		latitude
ψ		optical thickness

Subscript	Definition
a	air
b	bottom, bridge
c	conduction, convection, capillary
d	dew point, dry, damped

f	fluid
g	galvanometer
G	ground
Hg	mercury
i	ice, in
L	load
m	meter, mineral, mount, manometer
n	number, natural
o	out, organic matter, observed, surface level
p	parallel, plane
r	radiation, reference
s	shunt, surface
t	true, transient, top
T	temperature, transducer, thermistor
u	unknown
v	velocity, water vapor, vane
w	wall, weight, water, wet bulb
x	unknown
λ	wavelength
0	at zero °C, value at time zero

Chapter 1

Measurement Fundamentals

1.1   Introduction and Scope

Measurement programs should be planned with carefully defined objectives. Valid objectives include the verification of a hypothesis, the testing of a hypothesis, or explanation of phenomena. There is no place for measurement for the sake of measurement in a planned program. We hope that the techniques in this book will find their greatest usefulness in evaluating processes, such as growth, development, photosynthesis, or transpiration, rather than inventory or description of environmental factors.

The successful scientist must be capable of a sequence of activities that begins with a measurement program. First and foremost, the investigator should be an expert in the chosen field, with a thorough knowledge of the organisms or processes to be studied. Second, the investigator should know the instruments, their method of operation, and basic techniques for exposure and recording. Third, a knowledge of data analysis is required if the data are to be interpreted in terms of the objective. Calculators or computers are usually brought in at this step in order to analyze the data in terms of statistics, theory, and/or physical models. Finally, the observations, results, and conclusions should be reported to colleagues to avoid useless duplication of time and effort.

Mastery of the entire process normally comes after intensive training and a long period of experience. We shall focus on the second area: principles of instrumentation, exposure of instruments, and the recording of valid data. We will emphasize the validity of measurement rather than accuracy, as it is possible to accurately measure a temperature that is completely unrelated to the true value. We will bring our experience to bear on the problem of measuring true values of the desired entity.

1.2   Measurement Errors

Every measurement can be described with respect to accuracy, precision, and error. The definition of these terms at the outset will be helpful.

Accuracy is often confused with precision. *Accuracy* refers to the relation between the measured and "true" value, or the closeness to an accepted standard such as those maintained by the National Bureau of Standards. The true value plus the error is equal to the indicated value. *Precision*, on the other hand, refers to the variability observed among numerous measurements of a quantity. As an example, consider a micrometer that was initially both accurate and precise. If the micrometer is dropped and the frame bent, the accuracy is altered, but the precision would be unaffected if the lead screw remained undamaged. Accuracy is generally specified in terms of "inaccuracy." The accuracy of a thermometer, for example, may be accurate to $\pm 0.1^\circ$C over a given range.

The error may be composed of systematic and random components. A *systematic* error is unchanged between repeated measurements. For example, if a meter is not set to zero before making a series of measurements, the resulting errors would be consistently high or low. *Random* errors, in contrast will vary between measurements. They may be caused by such factors as electrical "noise," fluctuating temperatures, operator error, or wind. Many variables may contribute to random errors.

Random and systematic errors are illustrated in Fig. 1.1. The systematic error is the difference between the true value, V, and the mean of an infinite population of measurements, $\bar{Y}$. Random error is the difference between $\bar{Y}$ and the mean of a sample population, $\bar{X}$. As the sample size increases, the difference, $\bar{Y} - \bar{X}$, will decrease.

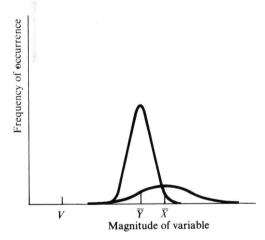

Figure 1.1 Illustration of systematic error $(V - \bar{Y})$ and random error $(\bar{Y} - \bar{X})$ where V is the true value, $\bar{Y}$ is the mean of an infinite number of measurements, and $\bar{X}$ is the mean of a sample population.

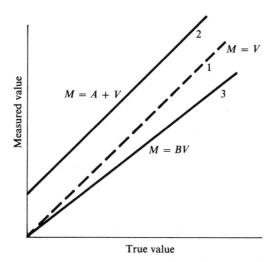

Figure 1.2. Various types of agreement between measured values, M, and true values, V, with intercept of A and slope of B.

Two types of systematic errors are illustrated in Fig. 1.2. Line 1 indicates perfect agreement between the measured and true values; line 2 differs from the true value by a constant amount; and line 3 differs by a constant slope, B. If there were additional data available for statistical analysis, the random error component could be illustrated by plotting confidence limits on either side of the lines.

The systematic and random error components can be added to indicate the range of error that may be expected in a specific reading. The error limits for a digital voltmeter, for example, may be given as $\pm(0.01\%$ of reading $+ 0.005\%$ of range $+ 1$ digit), indicating random components associated with the size of the measured value, the scale of the voltmeter, and the ambiguity of digital systems, respectively. The error limits will probably specify the conditions of measurement in order to exclude random errors associated with noise. If, for example, the voltmeter is reading a 60 mV signal with the range on 999.99 mV, the error limits would be

$$\pm(0.006 + 0.05 + 0.01) = \pm 0.066 \text{ mV}.$$

1.3   Estimating Error

Statistical techniques will yield the agreement between measured and predicted values when a number of observations are available, but it is often useful to estimate the error limits that may apply to a single measurement. The error is the difference between the measured value and the true value, and it may be expressed in units of measure, as a percentage, or as a unit error. From the voltmeter example given above, the unit error of the measurement would be $(0.066/60) = 1.1 \times 10^{-3}$.

The errors are not always available for a given instrument. For scale instruments the value may be estimated as $\pm 1/2$ of the smallest scale division. If, for example, a length of 6 cm is measured with a 10 cm ruler graduated in millimeter increments, the unit error would be $\pm(0.5 \text{ mm}/60 \text{ mm}) = 8.33 \times 10^{-3}$ of the reading. The errors are normally related to the measured value rather than the full scale reading.

1.3.1   Log Derivative Method

The errors in each of the measured variables can be combined to yield an estimate of total measurement error for a multiple variable problem. The log derivative approach can separate the contributions of each variable. Assume that A, B, C, D, F, and G are measured quantities, H is a constant, and E, the combined value of these quantities, is given by

$$E = \frac{HAB^4 C^{1/2}}{[D(F + G)]}. \tag{1.1}$$

To apply the log derivative method, first compute the unit error of each quantity, i.e., $\pm \Delta A/A$, $\pm \Delta B/B$, etc., from supplementary information provided with the instruments or the components in the system. Next, take the logarithm of the equation

$$\log E = \log H + \log A + 4 \log B + \tfrac{1}{2} \log C - \log D - \log(F + G). \tag{1.2}$$

Then differentiate the equation, recalling that the derivative of $d(\log y)/dy = dy/y$ and the derivative of a constant is zero;

$$\frac{dE}{E} = 0 + \frac{dA}{A} + \frac{4 \, dB}{B} + \frac{dC}{2C} - \frac{dD}{D} - \frac{d(F + G)}{F + G}, \tag{1.3}$$

$$\frac{dE}{E} = \frac{dA}{A} + \frac{4 \, dB}{B} + \frac{dC}{2c} - \frac{dD}{D} - \frac{dF}{(F + G)} - \frac{dG}{(F + G)}. \tag{1.4}$$

Rearranging and replacing the derivatives with finite differences (say $dE/E = \Delta E/E$) yields

$$\frac{\Delta E}{E} = \frac{\Delta A}{A} + \frac{4 \, \Delta B}{B} + \frac{\Delta C}{2C} - \frac{\Delta D}{\Delta} - \frac{\Delta F}{(F + G)} - \frac{\Delta G}{(F + G)}. \tag{1.5}$$

We can now substitute the unit errors calculated from supplementary information in the first step. In order for this method to apply, the unit errors should be less than 0.1. Since a unit error may be either positive or negative, we shall sum only absolute values to insure that the errors combine in the most unfavorable way.

The log derivative method is illustrated with the voltage divider circuit in Fig. 1.3. The problem is to estimate the errors in the output voltage, given information on the measuring meter and the circuit components. The 6.0 V

Figure 1.3. A simple voltage divider.

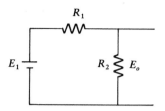

input voltage, E_1, is not affected by temperature. The 1 % precision resistors, R_1 and R_2, equal 10 kΩ and 100 Ω, respectively, and have a temperature coefficient of 50 ppm °C^{-1} from -55 °C to 165 °C.

In this example, both the input voltage, E_1, and the output voltage, E_0, are measured with a digital voltmeter whose rated accuracy is $\pm(0.01\%$ of reading $+0.005\%$ of range $+1$ digit). The voltmeter has a 5-digit readout and is set on a 10 V range (i.e., readout is 9.999 9 V). Using Ohm's law, the output voltage, E_0, can be predicted in terms of the input voltage and the resistances as

$$E_o = \frac{E_1 R_2}{(R_1 + R_2)}. \tag{1.6}$$

Taking logarithms, differentiating, and substituting yields

$$\log E_o = \log E_1 + \log R_2 - \log(R_1 + R_2), \tag{1.7}$$

$$\frac{dE_o}{E_o} = \frac{dE_1}{E_1} + \frac{dR_2}{R_2} - \frac{dR_1}{(R_1 + R_2)} - \frac{dR_2}{(R_1 + R_2)}, \tag{1.8}$$

$$\frac{\Delta E_o}{E_o} = \frac{\Delta E_1}{E_1} + \frac{\Delta R_2}{R_2} - \frac{\Delta R_1}{(R_1 + R_2)} - \frac{\Delta R_2}{(R_1 + R_2)}. \tag{1.9}$$

The unit error in the 6.0 V input voltage is

$$\frac{\Delta E_1}{E_1} = \frac{(0.6 \times 10^{-3} + 0.5 \times 10^{-3} + 0.1 \times 10^{-3})}{6} = 0.2 \times 10^{-3} \text{ V}. \tag{1.10}$$

The unit errors in R_1 and R_2 are one percent so $\Delta R_2/R_2 = 1/100 = 0.01$, $\Delta R_1/(R_1 + R_2) = 100/10\,100$, and $\Delta R_2/(R_1 + R_2) = 1/10\,100$. Summing the absolute value of all unit errors yields

$$\frac{\Delta E_o}{E_o} = 0.2 \times 10^{-3} + 10 \times 10^{-3} + 9.9 \times 10^{-3} + 99 \times 10^{-6}$$

$$= 20.2 \times 10^{-3}. \tag{1.11}$$

The output voltage for this circuit from Eq. (1.6) should be

$$E_o = \frac{6 \times 100}{10\,100} = 59.406 \text{ mV}. \tag{1.12}$$

For a signal of this magnitude, Eq. (1.11) says that the maximum error will be $\Delta E_o = 1.2$ mV from the uncertainties in the circuit components alone. Thus, the observed value would be 59.406 ± 1.2 mV.

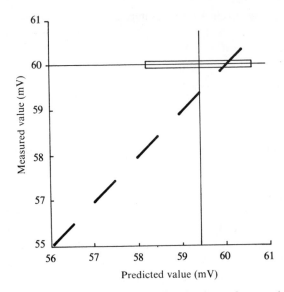

Figure 1.4. Comparison of measured and predicted values of a sample problem using Fig. 1.3.

Let us assume that the output measured with the voltmeter was 60 ± 0.066 mV. The measured and predicted values are shown in Fig. 1.4 with their areas of uncertainty. Agreement between the measured and predicted values is confirmed since a 1:1 line corresponding to perfect agreement, passes through the area of uncertainty.

It is advantageous to plot data with each division scaled to equal the value of the smallest division on the instrument or to the rated error, whichever is greater. In this case, the observed error of measurement is 0.066 mV and the smallest division on the voltmeter is 1 digit or 0.01 mV. The resolution of measurement is smaller than the error. Plotting the data on graph paper scaled to 0.06 mV will reduce distortion of the data due to overexpanded scales.

The error analysis can be extended to evaluate the effect of temperature upon the measurement. Using accurate resistors, divide Eq. (1.9) by a temperature increment $\Delta\theta$, a small change in temperature;

$$\frac{\Delta E_o}{E_o \Delta\theta} = \frac{\Delta E_1}{E_1 \Delta\theta} + \frac{\Delta R_2}{R_2 \Delta\theta} - \frac{\Delta R_1}{(R_1 + R_2)\Delta\theta} - \frac{\Delta R_2}{(R_1 + R_2)\Delta\theta} \quad (1.13)$$

Next, maintaining the signs and substituting the resistor specifications of $\Delta/\Delta\theta = 50 \times 10^{-6}\,°C^{-1}$ yields an estimate for the temperature coefficient of the output voltage of

$$\frac{\Delta E_o}{E_o \Delta\theta} = 0 + \frac{50 \times 10^{-6}(100)}{°C(100)} + \frac{50 \times 10^{-6}(1\,000)}{°C(10\,100)} - \frac{50 \times 10^{-6}(100)}{°C(10\,100)}.$$

$$(1.14)$$

In this case we maintain polarities when summing the unit errors as the temperature coefficients are fixed in operation. This circuit has several temperature sensitive elements with opposing effects that tend to cancel;

$$\frac{\Delta E_o}{E_o \, \Delta\theta} = (50 \times 10^{-6} - 49.5 \times 10^{-6} - 0.495 \times 10^{-6})°\text{C}^{-1}$$

$$= 5.0 \times 10^{-9}°\text{C}^{-1} \tag{1.15}$$

or

$$\frac{\Delta E_o}{\Delta\theta} = (5.0 \times 10^{-9}°\text{C}^{-1})(59.406 \text{ mV}) = 0.297 \times 10^{-6} \text{ mV } °\text{C}^{-1}. \tag{1.16}$$

The log derivative error analysis can be applied in another example to the well-known Stefan–Boltzmann law

$$E\uparrow = \varepsilon\sigma T^4, \tag{1.17}$$

where $E\uparrow$ is longwave radiation emitted from a surface of emissivity, ε, and temperature, T, given the Stefan–Boltzmann constant σ. Again, take the logarithm of the equation and differentiate to get

$$\frac{\Delta E\uparrow}{E\uparrow} = \frac{\Delta\varepsilon}{\varepsilon} + \frac{\Delta\sigma}{\sigma} + \frac{4\,\Delta T}{T}. \tag{1.18}$$

The result can illustrate several different assessments of errors. The basic equation is sometimes used with measurements of surface temperature and emissivity to estimate the emitted longwave radiation. It is also used with measurements of longwave radiation and emissivity to estimate surface temperature.

Let us consider first the value of the constant σ. It is prudent to assign some degree of uncertainty to constants, which could range as high as 0.1 for engineering constants unless specific information is available. The Stefan–Boltzmann constant is known rather precisely, however, to about $\pm 0.051\%$ based on three standard deviations [$\sigma = (56.697 \pm 0.029) \times 10^{-9}$ W m^{-2} K^{-4}, Page and Vigoureux, 1972]. For the purposes of this illustration, the error in the constant is negligible ($\Delta\sigma/\sigma \to 0$).

We can now consider errors in applying the equation to estimates of emitted longwave radiation. Estimates of emissivity are frequently in error. If we know the surface temperature precisely ($\Delta T/T = 0$), then Eq. (1.19) tells us that any error in estimating longwave radiation is directly proportional to errors in the emissivity,

$$\frac{\Delta E\uparrow}{E\uparrow} = \frac{\Delta\varepsilon}{\varepsilon} + 0 + 0. \tag{1.19}$$

On the other hand, we may seek to estimate surface temperature from measurements of longwave radiation. If we measure longwave radiation precisely ($\Delta E{\uparrow}/E{\uparrow} \to 0$), then

$$\frac{\Delta T}{T} = -\frac{1}{4}\frac{\Delta\varepsilon}{\varepsilon}, \tag{1.20}$$

and the unit error $\Delta T/T$ is one-quarter the unit error $\Delta\varepsilon/\varepsilon$. If $\Delta\varepsilon/\varepsilon$ is wrong by ± 0.1, the unit error in temperature will be 0.025. This is equivalent to an absolute error of ± 7.5 K at $T = 300$ K. This may be a larger error than the experiment can tolerate. Even so, it is apparent that emissivity must be known more precisely when estimating emitted longwave radiation than estimating surface temperature.

1.3.2 Probable Error Analysis

This generalized technique for evaluation of the uncertainties of measurement is based on methods outlined by Scarborough (1966). Application of these methods to environmental measurement problems has been made by Sinclair et al. (1975).

The general formula for the absolute error, ΔX, in a function X that consists of several variables y_i, each with its own absolute error, Δy_i, is

$$\Delta X = \frac{\Delta y_1}{\partial y_1}\frac{\partial X}{} + \frac{\Delta y_2}{\partial y_2}\frac{\partial X}{} + \cdots + \frac{\Delta y_n}{\partial y_n}\frac{\partial X}{}. \tag{1.21}$$

The general formula is equivalent to the total differentiation of the function X. The log derivative simplification discussed in Sect. 1.3.1 yields essentially the same result as the general formula in Eq. (1.21).

The magnitudes of the errors y_i can be expressed as absolute values of the worst possible case. This was the approach taken in the log derivative method in the preceding section in which the error is equivalent to the range of the largest possible errors in y_i. It is more likely that the errors about y_i are normally distributed, and there is some probability that errors in the different variables will tend to compensate each other to a limited degree. In this case, the formula for the probable error in X, δX, can be estimated by combining the individual errors through a least squares approach (Scarborough, 1966) that yields a formula similar to Eq. (1.21);

$$\delta X = \left[\left(\frac{\delta y_1}{\partial y_1}\frac{\partial X}{}\right)^2 + \left(\frac{\delta y_2}{\partial y_2}\frac{\partial X}{}\right)^2 + \cdots + \left(\frac{\delta y_n}{\partial y_n}\frac{\partial X}{}\right)^2\right]^{1/2} \tag{1.22}$$

To apply the probable error method, one needs only to differentiate the model in question with respect to the separate variables, multiply each differential by the error in that variable (known, measured, or estimated), and take the square root of the sum of the squared products.

As an example of this technique, let us estimate the probable error in the voltage divider of Eq. (1.6), and compare the results against the absolute error that was estimated by the log derivative method. Applying Eq. (1.22), we obtain

$$\delta E_o = \left[\left(\frac{\delta E_1}{} \frac{\partial E_o}{\partial E_1}\right)^2 + \left(\frac{\delta R_1}{} \frac{\partial E_o}{\partial R_1}\right)^2 + \left(\frac{\delta R_2}{} \frac{\partial E_o}{\partial R_2}\right)^2\right]^{1/2}$$

$$= \left[\left(\frac{\delta E_1 R_2}{(R_1 + R_2)}\right)^2 + \left(\frac{\delta R_1(-E_1 R_2)}{(R_1 + R_2)^2}\right)^2 + \left(\frac{\delta R_2 E_1 R_1}{(R_1 + R_2)^2}\right)^2\right]^{1/2}$$

$$(1.23)$$

Using the values from the earlier example, we have $E_1 = 6$ V and $\delta E_1 = 1.2$ mV [from the numerator in Eq. (1.10)]; $R_1 = 10\,000\ \Omega$, $R_2 = 100\ \Omega$, $\delta R_1 = 100\ \Omega$, and $\delta R_2 = 1\ \Omega$. The probable error in this example is

$$\delta E_o = [(11.88 \times 10^{-6})^2 + (0.588 \times 10^{-3})^2 + (0.588 \times 10^{-3})^2]^{1/2}$$

$$= 0.8319\text{ mV}. \tag{1.24}$$

The log derivative method [Eqs. (1.11) and (1.12)] estimates an absolute error of

$$\Delta E_o = (20.2 \times 10^{-3})(59.406) = 1.2\text{ mV}.$$

The probable error estimate is smaller than that of the absolute error, since some mutual compensation will occur if the errors are normally distributed. If we were to sum the absolute values of the errors listed in Eq. (1.23) in the same fashion as in the log derivative method, we find that the estimate of absolute error is 1.188 mV, as compared to $\Delta E_o = 1.2$ mV for the log derivative above. This confirms the equivalence of the log derivative and the total differential model [Eq. (1.21)].

Another technique for evaluating possible errors in X, given an error in one or more of the y_i's, is to simply increment the y_i variables by an appropriate estimate of error. If the increments are selected so as to produce the maximum possible error in the function X, the calculated error will be close to the maximum error value estimated by the log derivative method or the first general error model in Eq. (1.21). This technique is especially useful when the derivatives are difficult to evaluate. As an example, we can evaluate again the voltage divider model given in Eq. (1.6). First, we calculate the error-free output expected, as was done in Eq. (1.12), and obtain $E_o = 59.406$ mV. We next estimate $E_o + \Delta E_o$ after incrementing E_1 by $\Delta E_1 = 1.2$ mV, R_1 by $\Delta R_1 = -100\ \Omega$, and R_2 by $\Delta R_2 = 1\ \Omega$, and calculate

$$E_o + \Delta E_o = \frac{6.001 \times 101}{(9\,900 + 101)} = 60.6\text{ mV}.$$

After subtracting E_o, we obtain

$$\Delta E_o = 60.6\text{ mV} - E_o = 1.2\text{ mV}.$$

This is also the result obtained by the error estimation techniques of the log derivative and the general model for absolute error [Eq. (1.21)].

Table 1.1. Basic and derived units of the Système International with cgs and English equivalents.

Quantity	Dimension	SI	cgs	English
Length	L	1 m	10^2 cm	3.281 ft
Area	L^2	1 m^2	10^4 cm^2	10.76 ft^2
Volume	L^3	1 m^3	10^6 cm^3	35.31 ft^3
Time	T	1 s (or min, h)	1 s	0.2778×10^{-3} h
Velocity	$L T^{-1}$	1 m s^{-1}	10^2 cm s^{-1}	3.281 ft s^{-1}
Acceleration	$L T^{-2}$	1 m s^{-2}	10^2 cm s^{-2}	3.281 ft s^{-2}
Flow	$L^3 T^{-1}$	1 m^3 s^{-1}	10^6 cm^3 s^{-1}	127.1×10^3 ft^3 h^{-1} $= 951.1 \times 10^3$ gal h^{-1}
Mass	M	1 kg	10^3 g	2.205 lb
Density	$M L^{-3}$	1 kg m^{-3}	10^{-3} g cm^{-3}	6.24×10^{-2} lb ft^{-3}
Force	$M L T^{-2}$	1 kg m s^{-2} $= 1$ N (newton)	10^5 g cm $s^{-2} = 10^5$ dynes	0.224 lb f (lb force)
Pressure	$M L^{-1} T^{-2}$	1 kg m^{-1} s^{-2} $= 1$ Pa (pascal)	10 g cm^{-1} $s^{-2} = 10^{-2}$ mb	0.021 lb f ft^{-2}
Energy, work	$M L^2 T^{-2}$	1 kg m^2 s^{-2} $= 1$ J (joule)	10^7 g cm^2 $s^{-2} = 10^7$ erg	0.738 ft lb f
Power	$M L^2 T^{-3}$	1 kg m^2 s^{-3} $= 1$ W (watt)	10^7 g cm^2 $s^{-3} = 10^7$ erg s^{-1}	0.738 ft lb f s^{-1} $= 1.34 \times 10^{-3}$ hp (horsepower)
Temperature	θ	1 K (or 1°C)	1 K (or 1°C)	1.8°F
Energy, heat	H (or $M L^2 T^{-2}$)	1 J	0.2389 cal	0.948×10^{-3} BTU (British thermal unit)

Table 1.1. (*cont.*)

Quantity	Dimension	SI	cgs	English
Heat or radiation flux	$H\,T^{-1}$	1 W	0.2389 cal s^{-1}	3.414 BTU h^{-1}
Flux density or power per unit area	$H\,L^{-2}\,T^{-1}$	1 W m^{-2}	23.89×10^{-6} cal cm^{-2} s^{-1}	0.317 BTU ft^{-2} h^{-1}
Latent heat	$H\,M^{-1}$	1 J kg^{-1}	0.2389×10^{-3} cal g^{-1}	0.430×10^{-3} BTU lb^{-1}
Specific heat	$H\,M^{-1}\,\theta^{-1}$	1 J kg^{-1} K^{-1}	0.2389×10^{-3} cal g^{-1} °C^{-1}	0.2389×10^{-3} BTU lb^{-1} °F^{-1}
Thermal conductivity	$H\,L^{-1}\,\theta^{-1}\,T^{-1}$	1 W m^{-1} K^{-1}	2.389×10^{-3} cal cm^{-1} s^{-1} °C^{-1}	6.93 BTU ft^{-2} h^{-1} °F^{-1} in
Thermal diffusivity	$L^2\,T^{-1}$	1 m^2 s^{-1}	10^4 cm^2 s^{-1}	10.8 ft^2 s^{-1}
Dynamic viscosity	$M\,L^{-1}\,T^{-1}$	1 kg m^{-1} s^{-1}	10^5 g cm^{-1} s^{-1}	672 lb ft^{-1} s^{-1}
Kinematic viscosity	$L^2\,T^{-1}$	1 m^2 s^{-1}	10^4 cm^2 s^{-1}	10.8 ft^2 s^{-1}
Luminous intensity		cd (candela)		
Luminance		cd m^{-2}	10^{-4} cd cm^{-2}	0.0929 cd ft^{-2}
Luminous flux		1 cd sr = 1 lm (lumen)		
Illuminance		cd sr m^{-2} = 1 lx (lux)	10^{-4} cm^{-2} cd sr = 10^{-4} ph (phot)	0.0929 ft c (foot candle)
Electrical potential difference	$M\,L^2\,T^{-3}\,A^{-1}$	1 kg m^2 s^{-3} A^{-1} = 1 V		
Electrical charge	$A\,T$	1 A s = 1 C (coulomb)		
Electrical resistance	$M\,L^2\,T^{-3}\,A^{-2}$	1 kg m^2 s^{-3} A^{-2} = 1 Ω (ohm)		
Electrical capacitance	$M^{-1}\,L^{-2}\,T^4\,A^2$	1 kg^{-1} m^{-2} s^4 A^2 = 1 F (farad)		
Frequency	T^{-1}	Hz (hertz)		

1.4   Measurement Systems

Measurement systems provide the basis for all measurements; they associate specific unit values with various physical quantities. The Système International (SI) contains seven basic units defining length (meter, m), mass (kilogram, kg), time (seconds, s), electric current (ampere, A), temperature (kelvin, K), amount of substance (mole, mol), and luminous intensity (candela, cd). Two supplementary units apply to angular measurements— the plane angle (radian, rad) and the solid angle (steradian, sr). The SI system will be followed in this text.

The basic units and some derived units are given in Table 1.1 along with the cgs (centimeter, gram, second) and English equivalents. Prefixes and symbols used to form multiples are given in Table 1.2. In addition, some numerical values referred to in later chapters are listed in Table 1.3.

1.5   Significant Digits

Computers and electronic calculators can carry a large number of digits in the computation process. This frequently produces answers that appear to have much more precision than is warranted by the original input data. This is a particular problem when data of varying degrees of refinement must be combined in an analysis. Specific guidelines can be used when multiplying, dividing, adding, or subtracting to eliminate those digits that carry no information (ASTM 1972).

A digit that defines a specific value or quantity is significant. A zero may either indicate a specific value, as would any other digit, or merely the magnitude of a number. Consider a distance that was measured as 1 131.4 m. When rounded to the nearest meter and recorded as 1 131 m, it contains four significant digits. When rounded to the nearest 10 m and recorded as 1 130 m, it would contain three significant digits. If the distance were rounded to hundreds, it would be recorded as 1 100 m and would contain two significant

Table 1.2. Factors, prefixes, and symbols for forming multiples and submultiples in the SI system.

Factor by which the unit is multiplied	Prefix	Symbol	Factor by which the unit is multiplied	Prefix	Symbol
10^{12}	tera	T	10^{-2}	centi	c
10^{9}	giga	G	10^{-3}	milli	m
10^{6}	mega	M	10^{-6}	micro	μ
10^{3}	kilo	k	10^{-9}	nano	n
10^{2}	hecto	h	10^{-12}	pico	p
10	deka	da	10^{-15}	femto	f
10^{-1}	deci	d	10^{-18}	atto	a

digits. This example also illustrates some basic considerations in rounding. In all three cases the value of the right-hand digit was obtained by rounding from the measured value of an additional digit. The presence of zeros may thus simply indicate the magnitude of a rounded number. Each entry in the following list of numbers has one significant digit: 1 000; 100; 10; 0.01; 0.001.

Table 1.3. Some useful numerical values.

Physical constants	
Universal gas constant (R)	$8.314\,3$ J K^{-1} mol^{-1}
Boltzmann's constant (k)	13.81×10^{-24} J K^{-1} $molecule^{-1}$
Avogardro's number (N)	602.2×10^{21} mol^{-1}
Stefan–Boltzmann constant (σ)	56.696×10^{-9} W m^{-2} K^{-4}
Planck constant (h)	$0.662\,62 \times 10^{-33}$ J s
Velocity of light (c)	299.8×10^{6} m s^{-1}
Solar constant	1.38×10^{3} W m^{-2}
Wien's constant	$2\,897$ μm K
Acceleration due to gravity at 45° latitude and 0°C (g)	$9.806\,65$ m s^{-2}

Dry air	
Molecular weight of dry air (M_d)	28.97 g mol^{-1}
Density of air at 20°C and 1000 mb pressure (ρ)	1.209 kg m^{-3}
Specific heat at constant pressure (c_p)	$1\,004$ J K^{-1} kg^{-1}
Thermal conductivity at 20°C (k)	25.7×10^{-3} J $m^{-1}s^{-1}K^{-1}$
Dynamic viscosity at 20°C (μ)	18.18×10^{-6} kg $m^{-1}s^{-1}$
Kinematic viscosity at 20°C and 1000 mb (v)	15.29×10^{-6} m^2s^{-1}
Gas constant for dry air (R_d)	287 J $K^{-1}kg^{-1}$
Mechanical equivalent of heat	4.18 J cal^{-1}
Standard atmospheric pressure	101.3 kPa

Water substance	
Diffusivity of water vapor in air at 20°C and 1000 mb (O)	$25.7 \times 10^{-6}m^2s^{-1}$
Molecular weight of water (M_w)	18.016 g mol^{-1}
Gas constant for water vapor (R_v)	461 J K^{-1} kg^{-1}
Specific heat of water vapor at constant pressure	$1\,952$ J K^{-1} kg^{-1}
Specific heat of liquid water at 0°C	$4\,218$ J K^{-1} kg^{-1}
Latent heat of vaporization at 0°C (L)	2.50×10^{6} J kg^{-1}
Latent heat of fusion at 0°C	0.334×10^{6} J kg^{-1}
Thermodynamic psychrometric constant at 0°C (A)	0.646×10^{-3} K^{-1}

Rounding is the process of reducing a figure to fewer digits than the total number available. The basic rules are (1) leave the last digit retained unchanged if the first digit discarded is less than five, (2) increase the last digit retained by one unit if the first digit discarded is greater than five or if it is a five followed by at least one digit greater than zero, and (3) round the last digit retained to the nearest even number if the first digit discarded is five followed only by zeros. For example, 7.134 550 becomes 7.13 when rounded to three digits, 7.135 if rounded to four digits, and 7.134 6 if rounded to five digits.

The problems of retaining significance in computation can be minimized with a few simple guides. When adding or subtracting numbers, the answer should contain no more significant digits than is contained within the least accurate figure. For example, sum the distance measured earlier (1 131.4 m), rounded to units, tens, and hundreds of meters:

$$
\begin{array}{r}
1\ 131 \\
1\ 130 \\
1\ 100 \\
\hline
3\ 361
\end{array}
$$

The precision suggested by the total is exaggerated. In order to comply with the rule for adding and subtracting, each number should be rounded to one more right-handed significant digit than is contained in the least accurate number (1 100) as follows:

$$
\begin{array}{r}
1\ 130 \\
1\ 130 \\
1\ 100 \\
\hline
3\ 360
\end{array}
$$

The sum is then rounded to two significant digits, or 3 400, to put the final answer into a form compatible with precision of the input data.

The product or quotient must contain no more significant digits than are contained in the number with the fewest significant digits used in the multiplication or division. Consider, for example, that we wish to determine the velocity of an object that traversed the 1 131.4 m distance cited earlier in 203.1 s. The apparent velocity rounded to five decimal places would be 5.570 65 m s^{-1}. Other answers might be 5.568 69, 5.563 76, or 5.416 05 m s^{-1} depending upon whether the measured distance had been recorded with four, three, or two significant digits. The rule requires that the answers, rounded to the number of digits appropriate, would be 5.569, 5.56, or 5.4 m s^{-1}.

As another example, a typical calibration factor for a well-known radiometer is 12.3 μV W^{-1} m^2. If the radiometer output is read (accurately) to 9.367 mV, the radiation (9 367/12.3 = 761.544 7 W m^{-2}) should be recorded as merely 762 W m^{-2}.

An exact count is treated as if it contained an infinite number of digits. Thus, the product (or quotient) of a count and a measurement should contain the same number of significant digits as the measurement. If the distances in our earlier example (1 131, 1 130, 1 100 m) were to be divided by a count of ten, the acceptable answers would be 113.1, 113, and 110.

Bibliography

Brooks, P. P. B. (1962). Calibration procedures for direct-current resistance apparatus. Nat. Bur. Stand. U.S. Monogr. **39**.

Le Maraic, A. L., and J. P. Ciaramella (1975). The Metric Encyclopedia. Abbey Books, Somers, New York. 230 pp.

Luppold, David S. (1969). Precision of Measurements and Standards. Addison-Wesley, Reading, Mass. 251 pp.

Stein, Peter. (1967). Measurement Engineering. Stein Engineering Service, Phoenix, Arizona. 745 pp.

Literature Cited

American Society for Testing Materials (1972). Standard Metric Practice Guide (A Guide to the Use of SI—The International System of Units). American Society for

Page, C. H., and P. Vigoureux (eds.) (1972). The international system of units (SI). Nat. Bur. Stand. U.S. Spec. Publ. **330**.

Testing Materials, Philadelphia, Pa. 34 pp.

Scarborough, J. B. (1966). Numerical Mathematical Analysis. Johns Hopkins, Baltimore, Md. 6th ed.

Sinclair, T. R., L. H. Allen, Jr., and E. B. Lemon (1975). An analysis of errors in the calculation of energy flux densities above vegetation by a Bowen-ratio profile method. *Boundary-Layer Meteorol.* **8**:129–139.

Chapter 2

Review of Physical Fundamentals

The measurements discussed in this work deal with various aspects of the basic proportionality that exists between the transfer of an entity from one location to another, and to the potential for transfer that exists at various locations. The general, one-dimensional form of this relation is

$$\frac{dQ}{dt} = \frac{dA \; k \; dP}{dz},$$ (2.1)

where dQ/dt is the rate at which the entity Q is being transferred per unit of time, dA is the area through which the transfer occurs, k is a proportionality coefficient, and dP/dz is the change in potential for transfer with change in distance z.

The term dQ/dt is a rate of transfer or a flux. The flux becomes a flux density $(d^2Q/dt \; dA)$ when normalized per unit area by dividing Eq. (2.1) by dA. Some of the measurements to be discussed will deal with flux densities (radiant or conductive heat transfer), some will deal with the potential for transfer (voltage or temperature), and some will deal with the proportionality coefficient (electrical resistance).

Electric circuit components can be selected that will vary with changes in potential or flux density to produce corresponding changes in voltage potential, electric current, or electrical resistance. This interaction between the circuit elements and the desired environmental parameters provides the basic means for making environmental measurements. The basic relationships that exist between potentials and transfers will be discussed briefly.

2.1   Thermal and Latent Energy

Energy is defined in SI as the "work done when the point of application of a force of one newton is displaced through a distance of one meter in the direction of the force." The SI unit of energy is the joule (J). Work, energy, and quantity of heat are interchangeable concepts in SI, even though the joule is defined in terms of mechanical work. We shall restrict our considerations to thermal energy and to the latent energies of fusion or vaporization of water. These forms of energy are independent of position or of motion.

Latent energy is the energy released or absorbed as a substance changes phase (solid–liquid–gas). A specific amount of latent energy is associated with a specific mass of water, so the flux of water or water vapor is commonly considered in terms of the equivalent amount of latent energy. The latent heat of water is a slight function of temperature varying from 2.500 58 MJ kg^{-1} at 273.15 K to 2.256 67 MJ kg^{-1} at 373.15 K. The value at 293.15 K and 101.325 kPa pressure is 2.453 59 MJ kg^{-1}.

2.1.1   Potential for Transfer

The transfer potentials can be viewed as a way of specifying the energy levels at two different points. The actual thermal or latent flux density is related to the differences in energy level, as well as to the characteristics of the intervening medium.

One of the characteristics that affects the potential, and hence the transfer, is mass density, ρ, or unit mass; this is the mass per unit volume (kg m^{-3}). The specific volume v is the inverse of mass density ($\rho^{-1} = v$, m^3 kg^{-1}).

The mass density of dry air free of carbon dioxide is known precisely as a function of temperature and pressure. It is equal to (1.292 8 P)/(101.325) (1.000 028)(1 + 0.003 67 T), where air is at a pressure of P kPa and a temperature of T °C (1 kPa = 10 mb). Pure dry air at 101.325 kPa and 20°C has a density of 1.204 kg m^{-3} (Kaye and Laby, 1973). The mass density of air at 50% relative humidity and containing 0.04% carbon dioxide by volume is (1.293 07 P − 0.244 3 e_s)/(101.325)(1.000 028)(1 + 0.003 67 T) where P is atmospheric pressure in kPa, and e_s is saturated vapor pressure of water in kPa at air temperature, T, in °C. The density of air with 0.04% CO_2 and a relative humidity of 50% at 20°C and 101.325 kPa is 1.120 kg m^3. Corrections for other relative humidities can be found in Kaye and Laby (1973, p. 19).

The specific heat of a substance, c, is that quantity of heat required to raise its temperature one degree (K or °C). Units of c are J kg^{-1} K^{-1}. Temperature (K) is one of the fundamental dimensions in SI. The various temperature scales will be discussed in detail in Chapter 3. At this point, temperature represents the energy level or degree of thermal agitation of the molecules in a body. It indicates the potential for thermal energy transfer, but says nothing about the rate of transfer or the quantity of thermal energy available for transfer.

Specific heat is measured with respect to constant pressure, c_p, or constant volume, c_v. The specific heat of pure water at 15°C is 4 185 J kg^{-1} K^{-1}. It varies slightly with temperature T (°C) according to the relationship given by Kaye and Laby (1973),

$$\frac{c_p \text{ at } T \text{ (°C)}}{c_p \text{ at } 15°C} = 0.996\ 185 + 0.000\ 287\ 4\left(\frac{T + 100}{100}\right)^{5.26}$$

$$+ 0.011\ 160 \times 10^{-0.036\ T}. \tag{2.2}$$

The quantity of heat, Q, required to raise the temperature of a body of mass, m, from T_1 to T_2 is

$$Q = mc(T_2 - T_1). \tag{2.3}$$

The quantity of heat available for transfer between two temperature levels is thus governed by mass and specific heat of the substance in question.

2.1.2   Transfer Processes

The properties of the intervening medium have important effects on the transfer between two locations. We shall look at the different modes of transfer and specifically consider the transfer of thermal energy in a solid and in a vacuum, and the transfer of both thermal and latent energy in a gas.

2.1.2.1   Conduction Within Solids

Conduction is the process by which energy is transferred within a solid by mechanical interaction of adjacent molecules whose motions are the consequence of purely thermal agitation.

Equation (2.1) expresses the rate of conduction of heat in the normal direction in a homogenous medium when the potential P is that of temperature, T. On a steady-state basis, the quantity of heat conducted through area A in time, t, in the z direction is

$$Q = \frac{-A\lambda(T_2 - T_1)t}{(z_2 - z_1)}, \tag{2.4}$$

where T_2 and T_1 are the temperatures at z_2 and z_1. The negative sign ensures that heat will flow in the direction of a decreasing temperature gradient.

The thermal conductivity λ of a medium is the flux of thermal energy per unit area and unit temperature gradient. A related quantity, thermal diffusivity, is $k = \lambda/\rho\ c$. The SI units for conductivity are (W m^{-2}/K m^{-1} = W m^{-1} K^{-1}). The conductivity of many materials varies with temperature, and it is common practice to use the approximation $\lambda = (\lambda_2 + \lambda_1)/2$. The subscripts refer to the conductivity at T_2 and T_1, respectively. Numerical values of conductivity coefficients are available for many common materials in the various engineering handbooks and guides.

The conduction of heat radially through the walls of a pipe of length l with thickness $r_2 - r_1$ in the time interval t is given by

$$Q = \frac{-2\pi\lambda l(T_2 - T_1)t}{\ln(r_2/r_1)}. \tag{2.5}$$

Commonly, the conducting medium will consist of a number, n, of homogenous layers, each with a different thermal conductivity. The quantity of heat that is conducted through such a layered wall under steady-state conditions is

$$Q = \frac{-A(T_n - T_o)t}{\dfrac{z_1 - z_0}{\lambda_1} + \dfrac{z_2 - z_1}{\lambda_2} + \cdots + \dfrac{z_n - z_{n-1}}{\lambda_n}}, \tag{2.6}$$

where the temperature difference is $T_n - T_0$, the first layer thickness is $z_1 - z_0$, and the conductivity is λ_1, and the nth layer thickness is $z_n - z_{n-1}$ and the conductivity λ_n.

The quantity of heat that flows under steady state conditions through the walls of a multilayer pipe of length l in the time interval t is

$$Q = \frac{-2\pi l(T_n - T_0)t}{\dfrac{1}{\lambda_1}\ln\dfrac{r_1}{r_0} + \cdots + \dfrac{1}{\lambda_n}\ln\dfrac{r_n}{r_{n-1}}} \tag{2.7}$$

where $T_n - T_0$ is the temperature difference, $r_1 - r_0$ is the thickness of the first layer, and λ_1 is the conductivity of the first layer, and $r_n - r_{n-1}$ is the thickness and λ_n the conductivity of the nth layer.

2.1.2.2 Convection Within a Fluid

Convection is a mode of heat transfer during which thermal energy is transferred from one place to another by mass movement of parcels of the medium. This transfer process is thus restricted to substances that are fluid (liquid or gas). Convection is actually a special case of conduction, for the initial transfer of thermal energy at the interface between the solid and the fluid takes place by conduction. The distribution of this energy within the fluid is enhanced by the motion of the parcels of fluid operating either under heat-induced density differences (natural convection) or under the influence of outside forces (forced convection).

The basic transfer Eq. (2.1) can be rewritten for steady-state convection of heat across area A in time t as

$$Q = hA(T_f - T_w)t, \tag{2.8}$$

where T_f is the temperature of the fluid at some point removed from the influence of the convecting surface, T_w is the "temperature of the wall" that is transferring heat into the fluid, and h is the convection coefficient. The convection coefficient is the time rate of convection per unit area and per degree of temperature difference. The units of h are $W\ m^{-2}\ K^{-1}$. The numerical value varies with the static properties and the motion of the fluid (air, water, etc.). The convection between an object and fluid is also influenced by the shape of the object. Typical values for convection coefficients between an extended plane (a floor or wall) and still air would be about $5 < h < 10$ $W\ m^{-2}\ K^{-1}$; values with moving air would be in the order of $50 < h < 100$ $W\ m^{-2}\ K^{-1}$. A great deal of research has been conducted on the numerical values of convective heat transfer coefficients between fluids and objects of various shapes. The results are available in summary form in many engineering handbooks and guides.

Equation (2.8) applies only to the specific case in which convective energy is being exchanged between a surface and a fluid. A common problem in meteorology involves the estimation of energy convected between two points at a measured distance of separation within the atmosphere. This problem is evaluated with a formulation similar to Eq. (2.4) in which the distance $z_2 - z_1$ appears. The dimensions of the convection coefficient will thus differ from those used in the simplified Eq. (2.8).

The basic convection equation will also apply to the transfer of water vapor between an evaporating surface and some point in the atmosphere, or to the amount of vapor transferred between two points within the atmosphere. The flux density may be expressed in terms of the mass being transferred ($kg\ m^{-2}\ s^{-1}$), or in terms of energy, given the appropriate value of the latent heat of vaporization. The units of the convective transfer coefficient will depend upon the choice of units of concentration of water vapor. The various expressions for water vapor concentration will be discussed in detail in Chapter 6.

2.1.2.3   Transfer by Conduction and Convection Combined

Consider a rod whose base is attached to a wall with temperature T_w. The homogenous rod with uniform conductivity λ is long (with length l) with respect to its diameter d; it has a uniform cross-sectional area A and a uniform convective transfer coefficient $\bar{h}$ between the surface of the rod and the fluid. The rod protrudes into a fluid with uniform temperature T_f; the temperature gradient within the rod exists only in the z direction, so the problem reduces to a case of one-dimensional heat flow.

For the case where $T_w > T_f$, heat flows into the rod by conduction at the base and out of the rod by convection along its length. The temperature T diminishes along the length of the rod until at $z = l$, $T \to T_f$.

The general solution for temperature distribution in the rod in the z direction is

$$T - T_f = (T_w - T_f)e^{-mz}, \qquad (2.9)$$

where $m = 2(\bar{h}/\lambda d)^{1/2}$. The general solution of the total heat flow between the rod and the wall by conduction, or the rod and the fluid by convection, is

$$\frac{dQ}{dt} = -\left(\frac{\pi}{2}\right)(\bar{h}\, d^3\lambda)^{1/2}(T_w - T_f). \tag{2.10}$$

These are the basic forms of the solutions for temperature distribution and heat flow in a thermocouple wire inserted into a fluid medium of temperature different than that of the surroundings. The solutions match well with experimental results for rods whose lengths are very long when compared to the diameter.

More elaborate solutions are available for cases of short rods when the end effects become appreciable. Most engineering handbooks and heat transfer textbooks consider end effects and configurations other than cylindrical rods.

2.1.2.4   Transfer Through a Vacuum: Thermal Radiation

Radiation is the process by which thermal energy is transferred by electromagnetic waves from one point to another in the absence of an intervening medium. If an intervening medium is present, it must be at least partially transparent in order for the radiation transfer to take place.

Radiation has both direction and magnitude. The directional components can be important in certain measurement problems. These components can be examined with the aid of Fig. 2.1. Consider that the radiant flux or power (dQ/dt, W or J s^{-1}) arriving at an elemental surface area dA, is contained in an elemental cone that subtends a solid angle $d\Omega$. The cone is at an angle θ to the normal of dA. The radiant power received on this surface per unit area per unit solid angle is

$$\frac{d^3Q}{dt} = L\, dA\, d\Omega \cos\theta, \tag{2.11}$$

where the quantity L (J s^{-1} m^{-2} sr^{-1}) is known as the *radiance* of the field at dA in the direction of the cone. Note that a hemisphere contains 2π sr.

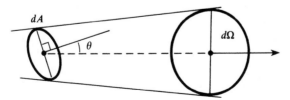

Figure 2.1. An elemental cone, $d\Omega$, that contains the radiant flux passing through an elemental area, dA, at an angle, θ, to the normal of dA.

The total radiant power falling from the hemisphere over one side of dA is given by integration of the solid angle over the entire hemisphere,

$$\frac{d^2Q}{dt} = dA \int_0^{2\pi} L \cos\theta \, d\Omega = E \, dA, \qquad (2.12)$$

where the quantity E (J s^{-1} m^{-2}) is a flux density known as the *irradiance*.

Equation (2.12) can be integrated when L is known as a function of θ and when L is independent of direction. When L is independent of direction the radiation field is *isotropic* and Eq. (2.12) integrates to $E = \pi L$. Thus, the flux density through an arbitrary surface in an isotropic radiation field is simply π times the radiance.

If the area dA is an element on the radiating surface, the energy lost through dA is the *emittance*. The energy lost by the whole of the radiating source is the *intensity* I (J s^{-1} sr^{-1}) defined by

$$\frac{d^2Q}{dt} = I \, d\Omega. \qquad (2.13)$$

Electromagnetic energy is defined by the wave nature of the radiation as well as by direction. The wave nature may be characterized by wavelength, λ (m), which is simply the distance from one wave crest to another, or by the frequency, v (Hz, s^{-1}), which is the rate at which wave crests pass a fixed point, or by the wave number, n (m^{-1}, $n = \lambda^{-1}$). These quantities are related by the relation $c = v\lambda = v n^{-1}$, where c is the speed of light ($c = 299.8 \times 10^6$ m s^{-1}). Spectral quantities are distinguished in various radiation equations by the subscripts λ or v, as appropriate.

Planck's law (Planck, 1914) relates the spectral emission to the temperature of the emitting body. His expression for the spectral radiance of a blackbody is

$$L_\lambda = c_1 \lambda^{-5} (e^{c_2/\lambda T} - 1)^{-1}, \qquad (2.14)$$

where $c_1 = 2hc^2 = 0.119\,11 \times 10^{-15}$ W m^2 sr^{-1}, and $c_2 = hc/k = 14.388 \times 10^{-3}$ m K. Planck's constant, h, is equal to $0.662\,62 \times 10^{-33}$ J s, and Boltzmann's constant, k, is equal to $13.806\,2 \times 10^{-24}$ J K^{-1}.

The irradiance or exitance of an isotropic radiation field is

$$E_\lambda = \pi L_\lambda = c_3 \lambda^{-5} (e^{c_2/\lambda T} - 1)^{-1} \qquad (2.15)$$

where $c_3 = \pi c_1 = 0.374\,18 \times 10^{-15}$ W m^2. The constants c_1 and c_2 above are the first and second radiation constants.

Calculation of the spectral flux density by means of Eq. (2.15) shows that there is a wavelength of maximum energy emission that varies with the temperature of the radiating body. Setting the derivative of Eq. (2.15) equal to zero and solving for the maximum point yields

$$\lambda_{max} = \frac{x}{T}, \qquad (2.16)$$

and when λ is in meters and T is in K, the constant x has a value of 2 897.9 μm K. This formulation of Wien's displacement law shows that the wavelength of maximum energy emission is inversely proportional to the absolute temperature of the radiation body.

The spectral flux density, or exitance, of Eq. (2.15) is plotted in Fig. 2.2 for temperatures associated with a range of activities at the surface of the earth. The dashed line in the figure illustrates the shift in wavelength of maximum energy emission (Wien's displacement law) as the temperature of the emitting body changes. The spectral exitance is tabulated in Table 2.1 at several wavelengths for the several temperatures illustrated in Fig. 2.2.

The total radiance over the spectrum is

$$L = \int_0^\infty L_\lambda \, d\lambda = \sigma' T^4, \tag{2.17}$$

where $\sigma' = 2\pi^4 k^4 (15c^2 h^3)^{-1} = 18.047 \times 10^{-9}$ W m^{-2} sr^{-1} K^{-4}, and the total exitance is

$$E = \sigma T^4, \tag{2.18}$$

where $\sigma = \pi\sigma' = 56.697 \times 10^{-9}$ W m^{-2} K^{-4} is known as the Stefan–Boltzmann constant.

A blackbody at any temperature radiates the maximum energy possible at that temperature while absorbing all incident radiation. Kirchoff's law

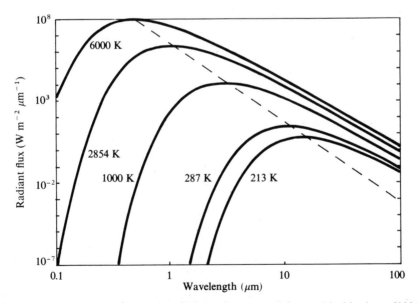

Figure 2.2. Spectral distribution of radiation flux density from a blackbody at 6000 K (sun), 2854 K (tungsten lamp), 1000 K (fire), 287 and 213 K (hot and cold earth). Dashed line represents wavelengths of peak emission according to Wien's law.

Table 2.1. Spectral radiation flux density $(W\,m^{-2}\,\mu m^{-1})$ for a black-body at various temperatures and wavelengths.

Wavelength	Temperature (K)				
(μm)	6 000	2 854	1 000	287	213
	$(\times 10^6)$	$(\times 10^6)$	$(\times 10^3)$		
0.2	7.27				
0.3	52.09	0.01			
0.4	91.35	0.12			
0.5	99.83	0.50			
0.6	90.13	1.08			
0.7	74.88	1.66			
0.8	60.01	2.10			
0.9	47.45	2.35			
1	37.42	2.44	0.21		
2	5.05	1.02	8.80		
3	1.26	0.35	12.84	0.08	
4	0.45	0.14	10.30	1.32	0.02
5	0.19	0.07	7.14	5.30	0.16
6	0.10	0.04	4.81	11.33	0.62
7	0.05	0.02	3.27	17.30	1.44
8	0.03	0.01	2.27	21.75	2.46
9	0.02	0.01	1.61	24.25	3.49
10			1.16	25.07	4.37
20			0.11	10.38	4.13
30			0.02	3.57	1.81
40			0.01	1.46	0.83
50				0.69	0.42
60				0.37	0.23

says that the emissivity and absorptivity is equal for all blackbodies at a given wavelength. This leads to a statement of the Stefan–Boltzmann law as

$$E = \varepsilon\sigma T^4, \qquad (2.19)$$

where ε is emissivity $(0 \leq \varepsilon \leq 1.0)$ of radiating body. This law provides the basic tool for estimating radiation transfer for black (or gray) bodies.

The values of ε vary with surface properties. Surfaces that are smooth with respect to the wavelengths of the emitted radiation have low emissivities (e.g., polished gold, 0.015; polished aluminum, 0.05). Surfaces that are rough with respect to the wavelength of the emitted radiation have high emissivities (e.g., paint, 0.90; surface of a leaf, 0.92). The emissivities of various substances are tabulated in engineering handbooks, but it is evident that no blackbodies exist in nature.

In general, an accurate solution of the thermal radiation transfer between two gray bodies can be obtained only if the bodies form a closed system. Such a system is approximated by two parallel surfaces that are large in extent. The quantity of radiation Q transferred between two infinite, parallel, gray planes of area $A(A_1 = A_2 = A)$ in time t is

$$Q = \bar{\varepsilon}\sigma A(T_1^4 - T_2^4)t, \tag{2.20}$$

where T_1 and T_2 are the Kelvin temperatures and $\bar{\varepsilon}$ is the mean emissivity (dimensionless) of the two surfaces. The mean emissivity is

$$\bar{\varepsilon} = \left(\frac{1}{\bar{\varepsilon}_1} + \frac{1}{\bar{\varepsilon}_2} - 1\right)^{-1}. \tag{2.21}$$

If the two gray surfaces are enclosed one within the other so that $A_1 \ll A_2$, the net transfer by radiation is

$$Q = \bar{f}\sigma A_1(T_1^4 - T_2^4)t, \tag{2.22}$$

where $\bar{f}$ (dimensionless) combines the effects of area and emissivity as

$$\bar{f} = \left[\frac{1}{\varepsilon_1} + \frac{A_1}{A_2}\left(\frac{1}{\varepsilon_2} - 1\right)\right]^{-1} = \left[\frac{1}{\bar{\varepsilon}_1} + \frac{A_1}{A_2}\left(\frac{1}{\bar{\varepsilon}_2} - 1\right)\right]^{-1}. \tag{2.23}$$

For very large A_2, Eq. (2.23) simplifies to $\bar{\varepsilon}_1$.

2.2 Basic dc Circuits

The concepts of potential and transfer apply directly to simple electrical circuits. Ohm's law relates the applied voltage, E, in a circuit to the product of the current, i, and resistance, R,

$$E = iR, \tag{2.24}$$

where E is the potential difference or the work done in moving one charge in the field of another charge. The unit of potential is called the volt, V, with one volt being equal to one joule per coulomb. Generally, the potential difference is expressed with respect to some reference level. Often this level is called the ground, which is considered to be an infinite electrical sink.

Current represents the movement of electrons or charge. Since the charge of an electron is 0.1603×10^{-18} coulombs, A s, current is thought of as the rate of movement of coulombs. Current is expressed in *ampere* A, where one ampere is equal to one coulomb per second.

The resistance of a circuit is the restriction to flow of current which is related to the electron field of the material used. Resistance is expressed in *ohms* Ω. One ohm is equal to one volt per ampere.

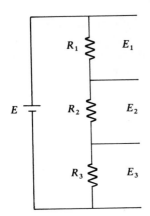

Figure 2.3. A series circuit.

Power, P, is used to express the dissipation of energy. It is the product of volts and amperes, or joules per second. Other useful expressions are

$$P = Ei = i^2 R = \frac{E^2}{R}. \qquad (2.25)$$

Resistors may be connected in parallel or in series giving rise to parallel, series, and combination circuits. The *series* circuit is illustrated in Fig. 2.3. The same current flows through all of the resistors since they are connected in series. However, the total voltage drop across the individual resistors is equal to the sum of the individual voltage drops,

$$E = E_1 + E_2 + E_3. \qquad (2.26)$$

By substitution,

$$E = iR_1 + iR_2 + iR_3, \qquad (2.27)$$

or

$$\frac{E}{i} = R_1 + R_2 + R_3 = \Sigma R. \qquad (2.28)$$

Assuming $R_1 = 20 \; \Omega$, $R_2 = 50 \; \Omega$, $R_3 = 30 \; \Omega$ and $E = 5$ V, we have

$$i = \frac{E}{\Sigma R} = \frac{5}{(20 + 50 + 30)} = 0.05 \text{ A}, \qquad (2.29)$$

and

$$E_1 = 0.05 \times 20 = 1 \text{ V}, \qquad (2.30)$$

$$E_2 = 0.05 \times 50 = 2.5 \text{ V}, \qquad (2.31)$$

$$E_3 = 0.05 \times 30 = 1.5 \text{ V}. \qquad (2.32)$$

Figure 2.4. A parallel circuit.

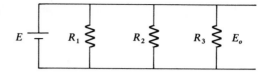

Contrasted to a series circuit, the voltage across all legs of a parallel circuit is the same and the total current is the sum of the individual currents. Consider Fig. 2.4,

$$i = i_1 + i_2 + i_3, \tag{2.33}$$

or

$$i = \frac{E_1}{R_1} + \frac{E_2}{R_2} + \frac{E_3}{R_3}. \tag{2.34}$$

Since $E_1 = E_2 = E_3 = E$,

$$i = \frac{E}{R_p} = E\left(\frac{1}{R_1} + \frac{1}{R_2} + \frac{1}{R_3}\right), \tag{2.35}$$

where R_p is the parallel resistance of the network. The equation for parallel resistance is

$$\frac{1}{R_p} = \frac{1}{R_1} + \frac{1}{R_2} + \frac{1}{R_3}. \tag{2.36}$$

When only two resistors are in parallel, Eq. (2.36) becomes

$$R_p = \frac{R_1 R_2}{R_1 + R_2} = \frac{(20 \times 50)}{20 + 50} = 14.3 \ \Omega. \tag{2.37}$$

The parallel resistance in Fig. 2.4 is

$$R_p = \frac{R_{p1} R_3}{R_{p1} + R_3} = \frac{(14.3 \times 30)}{14.3 + 30} = 9.7 \ \Omega. \tag{2.38}$$

The total current flow is

$$i = \frac{E}{R_p} = \frac{5}{9.7} = 0.52 \ \text{A}, \tag{2.39}$$

and $i_1 = 0.25$ A, $i_2 = 0.1$ A, and $i_3 = 0.17$ A.

A combination series and parallel circuit, sometimes called a loaded voltage divider, is used to illustrate one of the most common errors in environmental measurements, "the error of parallel resistors." Consider Fig. 2.5; if $R_1 = 400 \ \Omega$, $R_2 = 100 \ \Omega$, and R_3 is infinite, nonloaded output voltage E_o is $\frac{1}{5}$ of E or 10 mV if $E = 50$ mV. As the resistance of R_3 is decreased, the parallel resistance of R_2 and R_3 decreases with a resulting decrease in the loaded output voltage, E_L. The difference between E_L and E_o can be thought of as an output error and can be expressed as $100 (E_o - E_L)/E_o$

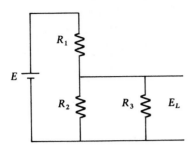

Figure 2.5. A combination circuit.

Table 2.2. The error associated with a loaded voltage divider.

$\dfrac{R_3}{\text{ohm}}$	$\dfrac{R_3}{R_2}$	$\dfrac{E_L}{\text{mV}}$	$100\,\dfrac{(E_o - E_L)}{E_o}$
1	0.010	0.123	98.765
10	0.100	1.111	88.889
100	1.000	5.556	44.444
1 000	10.000	9.259	7.407
10 000	100.000	9.921	0.794
100 000	1 000.000	9.992	0.080
1 000 000	10 000.000	9.999	0.008

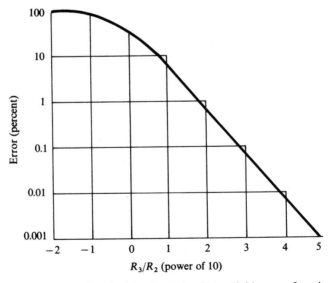

Figure 2.6. The error associated with a loaded voltage divider as a function of R_3/R_2 in Fig. 2.5.

for various values of R_3/R_2 (Table 2.2). The error decreases as the ratio R_3/R_2 increases. The error is approximately 0.1% when R_3/R_2 is 1000 (Fig. 2.6). This fact gives rise to a rule of thumb, "The impedance of a measuring device, e.g. R_3, must be at least 1000 times the resistance of the device being measured or the resistance of a voltage divider must be at least 1000 times the resistance of the transducer" (see Sect. 9.4.2).

2.3 Basic Measuring Instruments

2.3.1 Wheatstone Bridge

The measurement of a resistance requires an electric current to be passed through the element whose resistance is being measured. The resistance can be calculated from Ohm's Law relationships, based on the resistance element's effects on either current flow or potential. The measurements must minimize the effects of extraneous resistance changes (as may occur in the signal leads), unwanted variations in current, and self-heating caused by the current in the resistance element. Many specific circuits have been devised to facilitate application of a reference current and to develop the desired form of an output signal. The Wheatstone bridge is the basis for most of the circuits commonly used to measure resistance. Bridge circuits have high sensitivity and provide for a convenient comparison between unknown resistances and precisely known standards. All bridge designs include a set of standard resistances, a signal source, and an output indicator. The arrangement of these components depends on whether the bridge will be used in a null-balance or an unbalanced configuration.

The null-balance Wheatstone bridge is widely used to determine unknown resistances and temperature through temperature sensitive resistors. The schematic of an elementary bridge is shown in Fig. 2.7. The circuit consists of two fixed resistors, R_2 and R_3, one variable resistor R_1, an unknown resistor R_u, a galvanometer or null-detector, G, and a source of voltage, E.

The bridge can be considered as two parallel voltage dividers. Assuming R_2 and R_3 are of equal magnitude, the voltage drop across R_u is compared

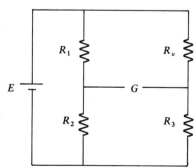

Figure 2.7. Wheatstone bridge. R_u is the temperature sensitive resistance element; R_1 is a variable resistor, R_2 and R_3 are fixed resistors, E is voltage applied, and G is a galvanometer.

with the voltage drop across R_1. If the voltage drops are equal, there is no potential difference between the midpoints of the bridge and current will not flow through the galvanometer when the switch is closed. The resistance of the unknown is equal to the resistance of R_1. The bridge is said to be in a state of null balance. If $R_u \neq R_1$, current will flow through the galvanometer and R_1 is adjusted, either increased or decreased, until a null balance is achieved.

At balance

$$\frac{iR_1}{iR_2} = \frac{iR_u}{iR_3}. \qquad (2.40)$$

Since the current passing through R_1 and R_2 is the same, i can be factored out. The current flowing through R_u and R_3 can be treated similarly. Consequently,

$$R_u = \frac{R_1 R_3}{R_2}. \qquad (2.41)$$

The ratio most often used in bridge applications is

$$\frac{R_2}{R_3} = \frac{R_1}{R_u}. \qquad (2.42)$$

From Eq. (2.42), one can conclude that if the ratio R_2/R_3 is known, then R_1 is the only value that must be known to determine R_u. For example, if $R_2/R_3 = 100$ and $R_1 = 100\ \Omega$, then R_u will be $1\ \Omega$.

When the bridge is operated in a balanced mode and a galvanometer is the null detector, the voltage required across the Wheatstone is given by

$$E = \frac{i\Sigma R}{P} \left(1 + \frac{R_g}{R_b} \right), \qquad (2.43)$$

where i is the necessary current in amperes to produce a detectable galvanometer deflection, ΣR is the sum of the resistance of the bridge arms, P is the precision required (expressed as a fraction), R_g is the galvanometer resistance, and R_b is the resistance of the bridge as viewed by the galvanometer.

The basic Wheatstone circuit is widely used for ac and dc bridges and a wide variety of commercial models are available. These self-contained units include the ratio arms, a variable resistance standard, keys for the galvanometer and power supply, and a suitable galvanometer. Well-designed bridges should be able to measure from about $0.1\ \Omega$ to the low M Ω range with about one percent accuracy. Other designs have advantages in certain applications. The Kelvin bridge, for example, is more accurate for low resistances.

An example of a general purpose null-balance bridge is shown in Fig. 2.8. R_1 is adjustable from 0.1 to $10\,000\ \Omega$, and the ratio of R_2/R_3 can be selected from 0.001 to 1 000.

The unbalanced bridge is preferred for use with automatic data logging

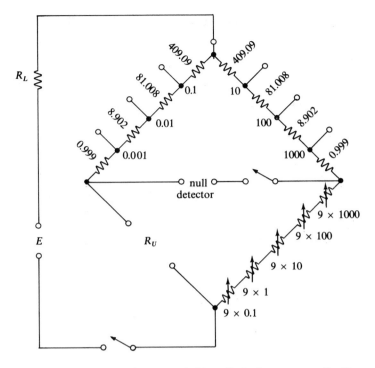

Figure 2.8. General purpose Wheatstone bridge. R_L is the current adjusting resistor, R_u is the unknown resistor, and E is the applied voltage. The resistances are given on the outside of the bridge and the resistance ratios are given on the inside of the bridge.

systems capable of measuring voltages. The unbalance can be detected as a potential difference or as a current between the center points of the bridge.

For design purposes, the basic bridge, Fig. 2.7, may be considered as two parallel voltage dividers—the unknown resistance divider, represented by $R_u + R_3$, and the zero reference divider represented by $R_1 + R_2$.

The values of the resistances R_1, R_2, and R_3 can be determined from the following analysis. Assume that a temperature bridge is desired with a range of 0°–50°C. The resistance of the temperature sensitive resistor, R_u, at 0°C is a, at 50°C is b, and at midrange, 25°C, is c. The voltage drop across R_u at 0°C is

$$E_1 = \frac{aE}{(a + R_3)}. \tag{2.44}$$

Similarly the voltage drop at 50°C is

$$E_2 = \frac{bE}{(b + R_3)}. \tag{2.45}$$

The maximum difference in voltage drop from 50° to 0°C is

$$\Delta E_o = E_2 - E_1 = \frac{bE}{(b + R_3)} - \frac{aE}{(a + R_3)}. \tag{2.46}$$

This equation has two unknowns, so a second equation expressing one-half range voltage drop is necessary for solution;

$$\tfrac{1}{2} \Delta E_o = \frac{bE}{(b + R_3)} - \frac{cE}{(c + R_3)}. \tag{2.47}$$

The unknown resistor, R_3, is determined by combining Eqs. 2.46 and 2.47;

$$R_3 = \frac{(bc + ac - 2ab)}{(a - 2c + b)}. \tag{2.48}$$

The applied voltage E can be expressed in terms of the maximum voltage out;

$$E = \frac{\Delta E_o(R_3 + a)(R_3 + b)}{[a(R_3 + b) - b(R_3 + a)]}. \tag{2.49}$$

The maximum power will be dissipated in R_u when its resistance is equal to R_3. Since power is $P = E^2/R$, the maximum applied voltage is $E = (PR)^{1/2}$. The power that can be dissipated by R_u is usually given by the manufacturer of the temperature sensitive resistor.

For maximum linearity, R_1 should equal R_u at the midrange, c. However, the output of the bridge would be zero at 25°C rather than at 0°C. For zero output at 0°C, R_1 should equal a. The bridge has been designed for perfect agreement between voltage out and temperature at three temperatures; 0°, 25°, and 50°C. The degree of nonlinearity at other temperatures depends on the linearity of R_u versus temperature.

A meter circuit, consisting of a meter, R_m, and a series resistor R_s may be substituted for G in Fig. 2.7. The meter circuit should have at least ten times the resistance of R_u at its lowest value. The voltage across the meter circuit is given by Eq. (2.49). The current flowing through the meter is

$$i = \frac{\Delta E_o}{10R_u}. \tag{2.50}$$

The meter used should have a full range value about equal to i. The resistance of the meter circuit is given by

$$R_m + R_s = \frac{(E/i)(R_1 R_3 - bR_2) - bR_1(R_2 + R_3) - R_2 R_3(b + R_1)}{(R_1 + R_2)(b + R_3)}. \tag{2.51}$$

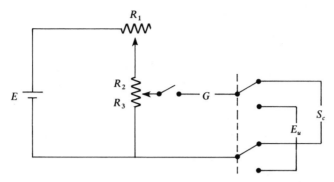

Figure 2.9. The basic potentiometer with symbols defined as the working voltage E, adjustable resistor R_1, voltage divider $R_2 + R_3$, galvanometer G, standard voltage S_C, and unknown voltage E_u.

The bridge is checked by (1) trimming R so that the current through the meter is zero when $R_u = a$, (2) adjusting R_L so maximum current deflection is indicated when $R_u = b$, and (3) checking to insure that the meter reads midrange at $R_u = c$.

2.3.2   Potentiometer

Most environmental sensors generate an emf or a potential. The magnitude of the emf or potential can be determined with an instrument called a potentiometer by comparing the unknown with an accurately known potential.

The potentiometer may be considered as a precision voltage divider. An elementary form is shown in Fig. 2.9. With the voltage divider set $(R_2 + R_3)$ to its maximum value $(R_2 = 0)$, the working voltage is standardized by adjusting the current limiting resistor R_1 until the potential difference between the voltage divider and a known standard voltage, S_c, is zero. This state is detected by the galvanometer G. Once the working voltage has been standardized, the unknown potential is connected in place of the standard voltage. The unknown potential is compared to the voltage drop across the voltage divider. The voltage divider is adjusted until the two potentials are equal as detected by zero current through the galvanometer. The unknown potential is equal to the voltage drop across the precision divider.

The accuracy of the potentiometer is dependent on the accuracy of the standard voltage and the calibration of the voltage divider.

The measurement of potentials over a large range say from zero to five V dc requires more elaborate potentiometer design including the Kelvin potentiometer circuit and a voltage divider for the unknown as shown in Fig. 2.10.

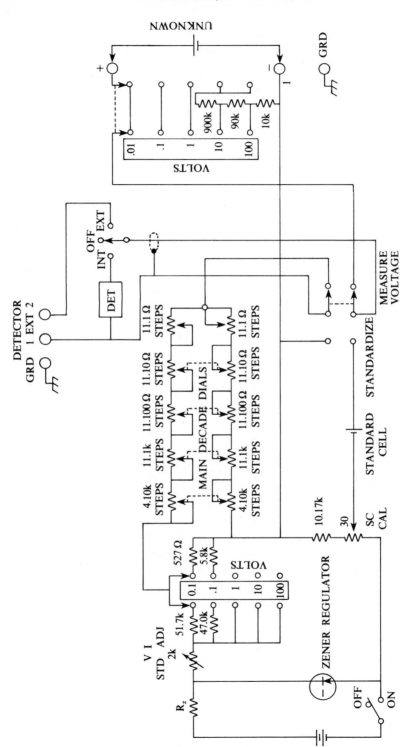

Figure 2.10. Potentiometric voltmeter circuit. Voltage standard adjust is VI STD ADJ, ground is GRD, internal is INT and external is EXT (Courtesy of Electro Scientific Industries).

Bibliography

Diefenderfer, J. A. (1972). Principles of Electronic Instrumentation. Saunders, Philadelphia, Pa. 675 pp.

Hickey, H. V. and W. M. Villines (1970). Elements of Electronics. McGraw-Hill, New York. 709 pp.

Kaye, G. W. C. and T. H. Laby (1973). Tables of Physical and Chemical Constants, 14th ed. Longmans, London. 386 pp.

Malmstadt, H. V., C. G. Enke, and E. C. Toren (1963). Electronics for Scientists. W. A. Benjamin, New York. 619 pp.

Planck, M. (1914). The Theory of Heat Radiation, 2nd ed. Translated by Morton Masius. Reprinted in 1959 by Dover, New York. 224 pp.

Kreith, F. (1973). Principles of Heat Transfer. Harper and Row, New York. 656 pp.

Chapter 3

Temperature

The temperature concept indexes the thermal energy level of a body. James Maxwell stated that "the temperature of a body is its thermal state considered with reference to the ability to communicate heat to other bodies." Note that Maxwell refers to the ability to transfer heat, rather than to the quantity of heat that may be transferred. The quantity of thermal energy gain (or loss) needed to cause a unit change in temperature depends on the mass of the body and the specific heat. Specific heat is the amount of thermal energy required to change the temperature of a unit mass by one temperature scale unit. Thus, for example, much more energy must be exchanged to create an equal change in temperature in a unit volume of water than in air.

3.1   Temperature Scales

The thermal energy level of an object is qualitatively interpreted with such descriptive terms as ice cold, warm, and boiling hot, but various temperature scales have been proposed in an attempt to quantify the terms. The scales that have gained acceptance use physical phenomena such as a change in phase from solid to liquid or liquid to gas. The ice and boiling points of water originally served as the standard reference points for temperatures near ambient. In terms of the familiar Celsius (°C) and Fahrenheit (°F) scales, the temperatures are close to 0°C (32°F) and 100°C (212°F). Thus, these two temperature scales are related in the following way: $T(°F) = 32 + \frac{9}{5} T(°C)$ and $T(°C) = \frac{5}{9} [T(°F) - 32]$, or in terms of differences, $1°C = 1.8°F$. The relationships between the several common temperature scales at reference points near ambient are summarized in Table 3.1.

The ice point has been superseded as a standard by the triple point of water. The triple point is the equilibrium state where ice, air-saturated water, and vapor-saturated air coexist. Its value is 0.01°C at 6.106 kPa. The triple point is also the defining equilibrium state for the thermodynamic temperature scale whose basic unit of one degree kelvin (1 K) is precisely 1/273.16 of the triple point of water. The thermodynamic temperature unit of the kelvin (K) honors William Thompson (Lord Kelvin) for his development of an absolute thermodynamic temperature scale in 1848.

A number of revisions have resulted in changes between the values of the fixed reference points, the functions, and instruments used to interpolate between the points. In 1968 the International Committee on Weights and Measures, under authority granted by the Thirteenth General Conference on Weights and Measures, accepted a revised scale referred to as the International Practical Temperature Scale of 1968 (IPTS-68). An English version of the official French text was prepared by the National Physical Laboratory (1969).

The IPTS-68 scale designates the platinum resistance thermometer as the instrument to serve as a standard between 13.81 and 903.89 K (630.74°C), with specified functions relating temperature to the resistance of a strain-free, annealed, pure platinum resistance element. The platinum–10% rhodium/platinum thermocouple is the accepted sensor for the range 903.89 to 1 337.58 K, while the Planck law of radiation is used above 1 337.58 K (with c_2 specified as 0.014 388 m K). If calibration measurements are contemplated using a platinum resistance thermometer within the ranges covered by IPTS-68, reference should be made to detailed procedures and suggestions based on current practice of the National Physical Laboratory (1969).

The triple point of water is easily established with a relatively simple apparatus. The standard triple point cell is a sealed glass chamber, partially filled with pure water of the isotropic composition of sea water. The chamber is evacuated to a pressure of 6.106 kPa (4.58 mm Hg), which is the only pressure at which the triple point will exist. An axial well for thermometers protrudes into the chamber. Ice is created inside the sealed chamber by introducing a suitable refrigerant into the access well; powdered dry ice (solidified CO_2) is often used. The ice sheath is observed while thickening, for

Table 3.1. Comparison of several near-ambient temperature references with various temperature scales. Pressures are 101.325 kPa for the ice and steam points and 6.106 kPa for the triple point.

Scale	Symbol	Ice point	Triple point	Steam point
Thermodynamic	K	273.15	273.16	373.15
Celsius	°C	0	0.01	100
Fahrenheit	°F	32	32.018	212
Rankin	°R	491.670	491.688	671.670

the refrigerant must be removed before the ice becomes thick enough to crack the chamber. Once the freezing has stopped, the access well can be filled with alcohol to insure uniform heat transfer between the triple point apparatus and the test sensors. The triple point temperature will be achieved once the ice sheath melts enough to establish a film of water around the access well. The triple point temperature can be maintained constant to about 0.000 1 K for months before refreezing is required, if the entire chamber is placed within a larger ice bath (National Physical Laboratory, 1969).

The boiling point reference is now called the steam point. The precise determination of the steam point is difficult because of its variation with atmospheric pressure. Corrections have to be applied for deviations from 101.325 kPa. The measurements and corrections are facilitated with commercially available hypsometers or steam baths. The IPTS-68 temperature as a function of vapor pressure of water is given to an accuracy of 0.000 1 K for the range from 99.9 to 100.1°C by the equation

$$T(°C) = 100 + 28.021\ 6\left(\frac{P}{P_0} - 1\right) - 11.642\left(\frac{P}{P_0} - 1\right)^2 + 7.1\left(\frac{P}{P_0} - 1\right)^3,$$

$$(3.1)$$

where P_0 is 101.325 kPa (National Physical Laboratory, 1969).

3.2   Time Constant

Every sensor exposed to a new environment requires some finite time to reach equilibrium. This adjustment time is a consideration for all types of instruments. It is of particular importance in the measurement of temperature because of the speed of temperature fluctuations. We shall use the example of a temperature sensor in the development of the mathematical definition of the speed of response or the concept of an instrument's *time constant*.

When a sensor at temperature T is placed in a medium of a different temperature, say T_a, the temperature of the sensor approaches the new temperature asymtotically. Observation has established that the change in sensor temperature with time, dT/dt, is proportional to the difference in temperature between the element and the environment $(T - T_a)$,

$$\frac{dT}{dt} = -\frac{1}{\tau}(T - T_a), \qquad\qquad (3.2)$$

where the coefficient of proportionality, τ, is known as the *time constant* of the sensor. Its dimensions are those of time (s).

In order to solve Eq. (3.2), we must have some information about the environmental temperature T_a, describing how it changes with respect to time. Of the virtually unlimited number of patterns of environmental temperature fluctuation, we shall select three forms for analysis: a step change, a ramp or constant change, and a periodic or sinusoidal change.

Consider first the step change in temperature as shown in Fig. 3.1(a). At time zero, t_0, both the environment and the sensor are at temperature T_0. The environmental temperature changes instantaneously to T_a (a step change), and the temperature of the sensor begins to rise toward the new temperature T_a. If temperature T_a remains constant, Eq. (3.2) can be integrated after first separating the variables to yield

$$\int \frac{1}{T - T_a} dT = \int -\frac{1}{\tau} dt, \tag{3.3}$$

and

$$\ln(T - T_a) + C = -\frac{t}{\tau}, \tag{3.4}$$

so

$$T - T_a = Ce^{-t/\tau}. \tag{3.5}$$

The constant, C, can be evaluated from the boundary condition that at $t = 0$, $T = T_0$. The final solution for the step change in temperature is

$$T - T_a = (T_0 - T_a)e^{-t/\tau}. \tag{3.6}$$

When the value of t equals τ, the ratio $(T - T_a)/(T_0 - T_a)$ equals $1/e$, or approximately 0.368. The value of τ is thus defined as the time required for the ratio of the temperature change remaining $(T - T_a)$ to that of the total change $(T_0 - T_a)$ to be reduced to the value of $1/e$. The value of τ will be unaffected by the magnitudes of the temperature changes. It is apparent, however, that if a time constant of τ specifies that the adjustment remaining is $1/e$ or 0.368 of the total adjustment, then it also specifies that the adjustment completed must be $1 - 1/e$ or 0.632 of the total adjustment. The multiples of the time constant required to achieve adjustments of 0.632, 0.900, 0.990, 0.999 and 0.999 9 are shown in Fig. 3.1(a). The degree of attainment of thermal equilibrium following a step change in temperature is tabulated in Table 3.2.

Table 3.2. Relation between elapsed time and the degree of achievement of thermal equilibrium.

Elapsed time(t)	Adjustme remaining ($e^{-t/\tau}$)	Adjustment completed ($1 - e^{-t/\tau}$)
1.0τ	0.368	0.632
2.3τ	0.100	0.900
3.0τ	0.050	0.950
4.6τ	0.010	0.990
6.9τ	0.001	0.999
9.2τ	0.000 1	0.999 9

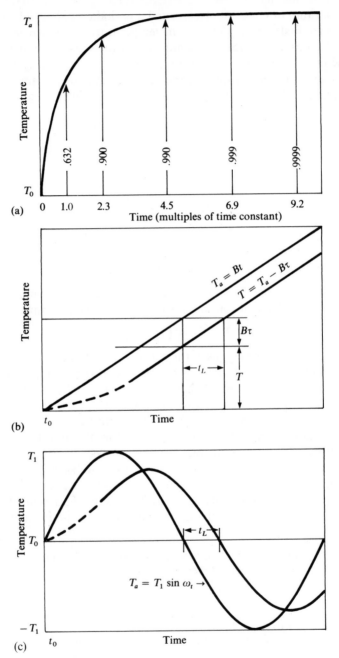

Figure 3.1(a)–3.1(c). Examples of time constants for (a) a step change, (b) a ramp change, and (c) a periodic change in the environmental temperature. For example, $\tau \simeq t_L$ in (b) and (c) whenever the elapsed time is much greater than τ.

The temperature difference between the sensor and the new temperature is reduced to five percent of the initial difference after a period of 3τ has elapsed and, for all practical purposes, thermal equilibrium is achieved after a period of 5τ has elapsed.

If the new temperature is changing at a constant rate, say $dT_a/dt = B$, we can express $T_a = Bt$, assuming $T = T_a = 0$ at the start. Eq. (3.2) becomes

$$\frac{dT}{dt} = -\frac{1}{\tau}(T - Bt), \tag{3.7a}$$

which can not be solved by simply separating the variables as in Eq. (3.2), but which can be rewritten as a linear, first-degree differential equation,

$$\frac{dT}{dt} + \frac{T}{\tau} = \frac{Bt}{\tau}, \tag{3.7b}$$

which has the general solution given in most mathematics texts of

$$T = B(t - \tau) + C_1 e^{-t/\tau}. \tag{3.7c}$$

If when $t = 0$, $T = T_0 = 0$, then $C_1 = B\tau$ and the final form of the solution is

$$T = Bt - B\tau(1 - e^{-t/\tau}). \tag{3.7d}$$

Since $T_a = Bt$, the equation can be written as

$$T - T_a = -B\tau(1 - e^{-t/\tau}), \tag{3.7e}$$

and when the elapsed time $t \gg \tau$, the solution reduces to

$$T - T_a = -B\tau. \tag{3.7f}$$

A temperature sensor lags behind a ramp change in temperature by an amount equal to the product of the time constant and the rate of temperature change, once the initial transient dies away. The response of the sensor to a ramp change is shown in Fig. 3.1(b).

In general, the environmental temperature is more closely represented by some periodic function than by a step or a ramp change. We can approximate the periodic function by a sine wave, varying from $+T_1$ to $-T_1$. Figure 3.1(c) illustrates the temperature oscillations $T_a = T_1 \sin \omega t$, where ω represents the angular frequency oscillation in radians per unit time and the amplitude T_1 is constant. Equation (3.2) becomes

$$\frac{dT}{dt} + \frac{T}{\tau} = \left(\frac{T_1}{\tau}\right)\sin \omega t, \tag{3.7g}$$

which is a linear, first-degree differential equation having the solution

$$T = C_1 e^{-t/\tau} + T_1[1 + (\omega\tau)^2]^{-1/2} \sin(\omega t - \arctan \omega\tau), \tag{3.7h}$$

when $T = T_a = 0$ initially. Evaluating the constant with $t = 0$ yields

$$T = T_1 \omega\tau[1 + (\omega\tau)^2]^{-1/2} e^{-t/\tau} + T_1[1 + (\omega\tau)^2]^{-1/2} \sin(\omega t - \arctan \omega\tau).$$
(3.7i)

Referring to Fig. 3.1c shows that the signal from the sensor is reduced in amplitude to the value αT_1, where $\alpha = [1 + (\omega\tau)^2]^{-1/2}$ is the attenuation factor. The signal from the sensor also lags behind the signal T_a by the amount arctan $\omega\tau$, once the elapsed time t exceeds τ by a substantial amount so that the transient term [the first term on the right-hand side of Eq. (3.7i)] approaches zero. The lag expressed as a phase angle (arctan $\omega\tau$) in radians can be converted to time units if divided by ω. To a close approximation, the time constant is the time that the sensor lags behind the environment once sufficient time has elapsed ($t \gg \tau$) so that the initial transient has died away [Fig. 3.1(c)].

The time constant of a sensor is related directly to the product of its heat capacity, ρc, and volume, V, and inversely to the ease with which heat can be transferred into the sensor (hA for convection, where h is the convective transfer coefficient and A is the area). For a system in which convection is the only mode of heat transfer to the sensor, and in which the sensor has infinite conductivity (no internal resistance to heat flow), the time constant is $\tau = \rho c V / hA$. This relation identifies the terms that affect the magnitude of the time constant. The time constant, for example, can be lengthened by increasing the heat capacity of a sensor or by decreasing the convective transfer coefficient while holding the volume/area ratio constant. Changes in volume and area, on the other hand, affect the heat capacity and convective transfer terms. However, the net effect of increasing the size of the sensor is to increase the time constant.

The time constant of a thermometer in air has been related empirically to wind speed, U, as $\tau = aU^b$, where a and b are constants that are associated with the physical dimensions and thermal characteristics of the thermometers. For standard mercury-in-glass thermometers with a spherical bulb, 1.12 cm in diameter, $a = 117$, $b = -0.48$, and $\tau = 56$ s for wind speeds of 4.6 m s^{-1} (Middleton and Spilhaus, 1960).

3.3　Measuring Devices

A thermometer is a transducer that converts temperature into an easily determined quantity. The thermometer output is based on the relationship between the thermal state of the thermometer and a physical property of the substance of which it is composed. Thermometers may be grouped into six classes according to their construction: (1) gas thermometers, (2) liquid-in-glass, (3) liquid-in-metal, (4) deformation, (5) electrical, and (6) sonic thermometers.

Table 3.3. Values of volume expansion at room temperature for commonly used thermometer materials.

Material	Volume expansion per °C ($\times 10^{-3}$)
Methyl alcohol	1.4
Ice	0.7–0.9
Water	0.21
Mercury	0.18
Glass	0.02–0.03

The gas thermometer is based on the principle that the pressure of a gas held at constant volume varies with temperature. This thermometer is very accurate and was at one time the standard for temperature measurements.

The liquid-in-glass, liquid-in-metal, and deformation thermometers utilize differing coefficients of thermal expansion for their readout. The coefficients for several different materials are tabulated in Table 3.3. Alcohol has a large expansion coefficient (45 to 80 times that of glass) and a low freezing point. Consequently, it is an excellent substance for thermometer construction. The accuracy of alcohol-filled thermometers, however, is less than with those filled with mercury, even though alcohol has a much larger coefficient of thermal expansion.

The deformation thermometers will not be discussed in detail because they are not adapted to electronic recording. They include the bimetallic strip and the Bourdon tube. The exposure errors common to all thermometers will be discussed, however.

Electric thermometers are adapted to automatic recording and will be discussed in detail. Basically, there are two main types of electrical transducers—self-generating and non-self-generating. The self-generating transducers produce an electric current as a function of temperature. The non-self-generating transducers require the application of an external signal in order to detect a change in property. Thermocouples are examples of the former while resistance elements, thermistors, and diodes are examples of the latter. The advantages and disadvantages of these types will be discussed in Section 3.3.3.

3.3.1 Thermocouples

Thermocouples have won a permanent role as temperature sensors for numerous industrial applications. Their favorable characteristics include acceptable accuracy, suitability over a wide range of temperatures, adequate thermal response, ruggedness, high reliability, low cost, ease of installation, and compatibility with most measuring and recording systems.

Thomas Johann Seebeck discovered in 1821 that a current would flow in a circuit composed of two dissimilar materials when the two junctions were at different temperatures. The direction of the current and the magnitude of the electromotive force (emf) produced depended on the materials and the temperature difference between the two junctions. The *Seebeck effect* is the conversion of thermal energy to electrical energy; it depends on the ease with which excess electrons will circulate in an electric circuit under the influence of temperature.

The emf generated by a combination of two materials, A and B, is proportional to the temperature difference between the junctions when the ends are connected to form a loop. A small change in temperature, dT, will generate a change in the voltage $d(\text{emf})$,

$$d(\text{emf}) = N_{A,B}\, dT, \tag{3.8a}$$

where the coefficient of proportionality, $N_{A,B}$, is called the *Seebeck coefficient* or the thermoelectric power of the combination A and B.

The *Peltier effect* was reported in 1834 by Jean Peltier who observed that the current in one direction across a junction of dissimilar materials would warm the junction and release heat. If the current were reversed, the junction would cool and absorb heat. The effect occurred regardless of whether the current came from an external source or was induced by differential heating (or cooling) of the thermocouple junctions. The effect was reversible, so the Peltier effect is closely related to the Seebeck effect. A *Peltier coefficient* can be developed for different materials in a manner analogous to the Seebeck coefficient by relating the generation of heat per unit of current.

William Thompson found in 1851 that a current flowing through a single homogenous conductor would absorb or liberate heat if a temperature gradient existed in the conductor. The *Thompson effect* was evident both in currents introduced from external sources and those generated by the thermocouple itself. The *Thompson coefficient* defines the ability of a given material to generate heat with respect to both a unit temperature gradient and a unit of current. The Thompson coefficient also represents the emf that will be generated in a homogenous conductor per unit of temperature gradient.

Benedict (1977) shows that the net Seebeck emf is composed of four distinct emfs. For materials A and B with junction temperatures T_1 and T_2, the net emf generated is the sum of the two Peltier effects (one at the T_1 and the other at the T_2 junction) and the two Thompson effects (one in A and the other in B) for the temperature difference $T_1 - T_2$. The Peltier effect and the Thompson effect can work to minimize the temperature between the two junctions. The effects are usually negligible, however, because the amount of heat evolved is quite small with respect to the amount of thermal energy available from the environment to the junctions at T_1 and T_2. Further, when potentiometric recording devices are used for measuring the emf, the current is negligible in the thermocouple circuit.

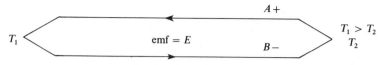

Figure 3.2. Seebeck effect.

The Seebeck effect is illustrated in Fig. 3.2, where junctions are formed between metals A and B. If one junction is at T_1 and the other junction is at a higher temperature T_2, a current will flow and will remain flowing as long as the temperature difference exists. The emf generated may be measured by inserting a voltmeter into either leg of the loop. A 0.56-mm copper–constantan thermocouple is shown in Fig. 3.3(a).

3.3.1.1 Thermoelectric Laws

Much experimentation with the thermocouple circuits has led to the formulation of three empirical laws which are fundamental to the accurate measurement of temperature by thermoelectric means.

The Law of Homogeneous Materials: "An electric current cannot be sustained in any circuit of a single homogeneous material, however varying in section, by the application of heat alone."

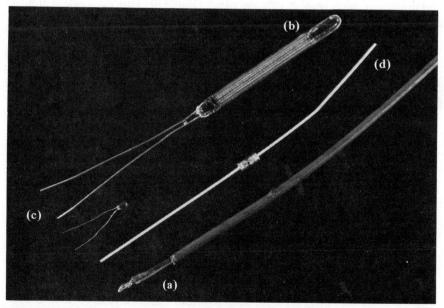

Figure 3.3. Temperature sensors: (a) 0.56-mm diameter copper–constantan thermocouple junction; (b) 100-Ω platinum resistance element, 3-mm diameter and 30-mm length; (c) 10 kΩ glass coated bead thermistor, 1.1-mm diameter; (d) glass encapulated silicon diode, 1.8-mm diameter and 4.2-mm length.

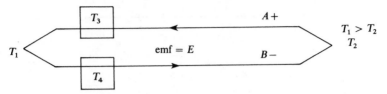

Figure 3.4. The Law of Homogenous Materials: T_3 and T_4 do not affect E.

As long as the metal of each wire in a thermocouple is homogeneous, the emf generated at the junction will not be affected by temperatures in the lead wire even though a temperature distribution exists along the lead wire, provided, of course, that the temperature is not conducted to the effective junction (Fig. 3.4).

The Law of Intermediate Materials: "The algebraic sum of the thermo-electromotive forces in any circuit composed of a number of dissimilar materials is zero if all of the circuit is at a uniform temperature."

Any measuring device or lead wire can be added to the circuit without affecting the accuracy as long as the new junctions are at the same temper-ature (Fig. 3.5). This also permits soldering or brazing of thermocouple junctions as the introduction of the binding material will not affect a junction that is at a uniform temperature.

Thermal emfs of two metals with respect to a third, C, may be added to form the emf with respect to each other (Fig. 3.6). This principle is used in calibration of various metals for thermocouple effects.

The Law of Successive or Intermediate Temperatures: "If two dissimilar homogeneous metals produce a thermal emf of E_1, when the junctions are at the temperatures T_1 and T_2, and a thermal emf of E_2 when the junctions are at temperatures T_2 and T_3, the thermal emf generated when the junctions are at temperatures T_1 and T_3 will be $E_1 + E_2$."

This law permits the determination of the measuring junction temperature from calibration charts based on a certain reference junction temperature (e.g., 0°C) when the reference junction is at a different but known temperature (e.g., 40°C) (Fig. 3.7).

By combining the three basic thermoelectric laws, it seems that (1) the

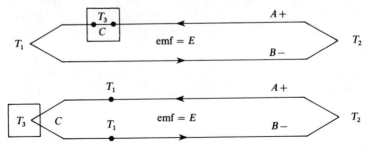

Figure 3.5. The Law of Intermediate Materials: third material, C, does not affect E.

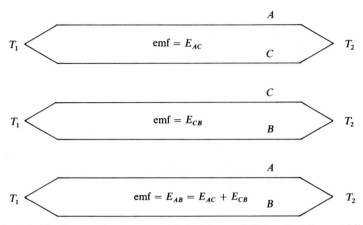

Figure 3.6. The Law of Intermediate Materials: emfs for materials are additive.

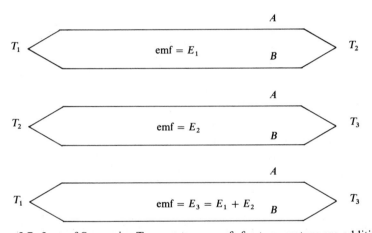

Figure 3.7. Law of Successive Temperatures: emfs for temperature are additive.

algebraic sum of the thermoelectric emf generated in any given circuit containing any number of dissimilar homogeneous metals is a function only of the temperatures of the junctions and (2) if all but one of the junctions in such a circuit are maintained at some reference temperature, the emf generated depends only on the temperature of that junction and can be used as a measure of its temperature.

3.3.1.2 Emf/Temperature Relationships

The Law of Intermediate Materials showed that the Seebeck emf of A with respect to C ($N_{A,C}$) and of B with respect to C ($N_{B,C}$) can be added to obtain that of A with respect to B;

$$N_{A,B} = N_{A,C} + N_{B,C}. \tag{3.8b}$$

Platinum is frequently used for a reference material, as it is relatively easy to obtain in a pure state. Determining the Seebeck coefficients with respect to platinum provides a basis for comparing the characteristics of materials over time and between manufacturers.

The Seebeck coefficients of four common thermocouple materials—iron $(J+)$, constantan $(J-)$, copper $(T+)$, and constantan $(T-)$—are tabulated in Table 3.4 with respect to platinum for each of the four materials. The sums yield Seebeck coefficients for the standard thermocouple combinations of iron–constantan (J) and copper–constantan (T). The platinum used is platinum-67, a reference standard maintained by the National Bureau of Standards; data is from the National Bureau of Standards Monograph 125 (Powell et al., 1974). The emf with respect to platinum of these and several other common thermoelectric metals are plotted in Fig. 3.8.

Note that the constantan used with iron $(J-)$ in Table 3.4 has a slightly different Seebeck coefficient than that used with copper $(T-)$. The composition of the two alloys differs slightly, and they should not be interchanged without prior calibration. The $T-$ constantan is often referred to as Adams constantan, and $J-$ as SAMA constantan.

The Table 3.4 and Fig. 3.8 show that various combinations of materials will give differing thermoelectric powers. A number of groupings of materials have been recognized for various industrial and scientific purposes over the years. The materials are commonly classed as noble metal or as base metal thermocouples. The noble metal combinations most frequently used include platinum–30% rhodium/platinum (type B), platinum–13% rhodium/platinum (type R), and platinum–10% rhodium/platinum (type S). The base metal thermocouples that are commonly encountered include chromel–

Table 3.4. Seebeck coefficients of common thermocouple metals with respect to platinum-67. Metals are iron $(J+)$, SAMA constantan $(J-)$, copper $(T+)$ and Adams constantan $(T-)$. The sum of the two J components yields the Seebeck coefficient for iron–constantan (type J) thermocouples. The sum of the two T components yields the Seebeck coefficient for copper–constantan (type T) thermocouples (Powell et al., 1974).

Temperature (°C)	Seebeck coefficients (μV °C^{-1})			
	$J+$	$J-$	$T+$	$T-$
0	17.9	32.5	5.9	32.9
50	17.9	34.9	7.8	35.0
100	17.2	37.2	9.4	37.4
150	16.0	39.2	10.6	39.5
200	14.6	40.9	11.9	41.3
250	13.1	42.4	13.1	42.7

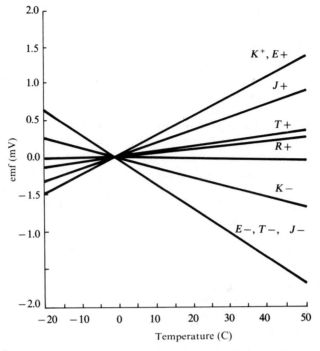

Figure 3.8. Emf versus temperature with respect to Platinum-67 for some common thermocouple materials. Platinum versus: chromel, $E+$, K^+; iron, $J+$; copper, $T+$; platinum–13% rhodium, $R+$; alumel, $K-$; constantan, $E-$, $T-$, $J-$. Data from National Bureau of Standards (Powell et al., 1974).

constantan (type E), iron–constantan (type J), chromel–alumel (type K), and copper–constantan (type T).

Note that the platinum–10% rhodium/platinum thermocouple is the standard instrument used to interpolate the IPTS-68 scale between the temperatures of 630.74° and 1064.43°C. The temperature (in °C) is defined by the second-degree equation

$$\text{emf} = a + bT + cT^2, \tag{3.8c}$$

where emf is the electromotive force of a standard thermocouple of platinum–10% rhodium alloy versus platinum, when one junction is at the temperature 0°C and the other is at the unknown temperature T °C. The constants a, b, and c are calculated from the values of emf at 630.74°C $\pm$ 0.2°C, as determined by a platinum resistance thermometer, and at the freezing points of silver and gold (National Physical Laboratory, 1969). The Seebeck coefficient of the type-S thermocouple, in keeping with the other noble metal thermocouples, is rather low, being only 6 μV°C^{-1} at 25°C, and ranging from 10.3 to 11.7 μV°C^{-1} over the IPTS-68 range that it defines.

The copper–constantan (type T) thermocouple is one of the oldest and most popular thermocouples for determining temperatures within the range from about 300°C down to the boiling point of hydrogen at 20.28 K. The Seebeck coefficients near ambient range from 38.7 $\mu V°C^{-1}$ at 0°C to 46.8 $\mu V°C^{-1}$ at 100°C. The limit of calibration error specified for type-T commercial thermocouple wire is ± 0.8°C between $-59°$ and 93°C; the limit for premium thermocouples wire over this range is ± 0.4°C. The copper $(T+)$ material minimizes offsets from parasitic emfs associated with the connection of recording instruments. However, the high conductivity of the copper requires that special care be taken to ensure that both the reference and the measuring junctions are at the appropriate temperature.

The type-T calibration curves developed by the National Bureau of Standards (Powell et al., 1974) use an eight-degree equation to fit 17 key data points between 0° and 400°C with an imprecision of 0.7 μV. A 14-degree equation was required to define the thermoelectric voltage between 0° and -270°C. The coefficients of these equations are tabulated in Table 3.5 to facilitate calculations. The table also contains a fourth-degree approximation over the same range with slightly greater error. The inverse of the fourth-degree approximation equation, used to estimate temperatures between 0° and 400°C from emf measurements, is tabulated in Table 3.6 along with similar approximations for other common thermocouples. The emf of type-T thermocouples with respect to a 0°C reference temperature is tabulated in Table 3.7 at intervals of 1°C.

The chromel–alumel (type K) thermocouple is resistant to oxidation and can be used at higher temperatures than types T or J. Type-K thermocouples will withstand continuous use at temperatures within the range $-250°$ to 1260°C. Chromel $(K+)$ is an alloy of about 90 % nickel, 9–9.5 % chromium, about 0.5 % silicon and iron, and traces of carbon, manganese, cobalt, and niobium. Alumel $(K-)$ is typically composed of 95–96 % nickel, 1–1.5 % silicon, 1–2.3 % aluminum, 1.6–3.2 % manganese, up to about 0.5 % cobalt, and traces of iron, copper, and lead (Powell et al., 1974). Both thermoelements have relatively low thermal conductivities. The Seebeck coefficients range from 39.5 $\mu V°C^{-1}$ at 0°C to 41.4 $\mu V°C^{-1}$ at 100°C. The thermoelements will match the standard emf tables between 0° and 277°C within ± 2.2°C; premium wire will match within ± 1.1°C. The National Bureau of Standards (Powell et al., 1974) has developed to ten-degree equation to cover the temperature range from $-270°$ up to 0°C, and an eight-degree equation for the range from 0° to 1 372°C. The emf of type-K thermocouples is tabulated for ten-degree increments in Table 3.8 for the range from $-100°$ to 1 390°C with respect to a 0°C reference. The coefficients for a fourth-power equation used to estimate temperature (°C) from emf (μV) are tabulated in Table 3.6.

The iron–constantan (type J) thermocouple is widely used by industry because of its low cost and relatively high Seebeck coefficient. There is a lack of uniformity and standardization of type-J components, so the materials from different manufacturers may deviate considerably from the standard

Table 3.5. Emf as a function of temperature for copper–constantan (type-T) thermocouples. The power series expansion is of the form $\text{emf} = \sum_1^n a_n T^n$, where emf is in microvolts and T in degrees Celsius (Powell et al., 1974).

	Range: −270° to 0°C		Range: 0° to 400°C	
	Precise	Approximate	Precise	Approximate
a_1	$3.874\ 077\ 384\ 0 \times 10^{1}$	$3.943\ 991\ 9 \times 10^{1}$	$3.874\ 077\ 384\ 0 \times 10^{1}$	$3.846\ 840\ 7 \times 10^{1}$
a_2	$4.412\ 393\ 248\ 2 \times 10^{-2}$	$6.240\ 745\ 2 \times 10^{-2}$	$3.319\ 019\ 809\ 2 \times 10^{-2}$	$4.665\ 173\ 1 \times 10^{-2}$
a_3	$1.140\ 523\ 849\ 8 \times 10^{-4}$	$8.077\ 356\ 8 \times 10^{-5}$	$2.071\ 418\ 364\ 5 \times 10^{-4}$	$-3.737\ 579\ 3 \times 10^{-5}$
a_4	$1.997\ 440\ 656\ 8 \times 10^{-5}$	$2.684\ 564\ 7 \times 10^{-7}$	$-2.194\ 583\ 482\ 3 \times 10^{-6}$	$1.599\ 983\ 3 \times 10^{-8}$
a_5	$9.044\ 540\ 118\ 7 \times 10^{-7}$		$1.103\ 190\ 055\ 0 \times 10^{-8}$	
a_6	$2.276\ 601\ 850\ 4 \times 10^{-8}$		$-3.092\ 758\ 189\ 8 \times 10^{-11}$	
a_7	$3.624\ 740\ 938\ 0 \times 10^{-10}$		$4.565\ 333\ 716\ 5 \times 10^{-14}$	
a_8	$3.864\ 892\ 420\ 1 \times 10^{-12}$		$-2.761\ 687\ 804\ 0 \times 10^{-17}$	
a_9	$2.829\ 867\ 851\ 9 \times 10^{-14}$			
a_{10}	$1.428\ 138\ 334\ 9 \times 10^{-16}$			
a_{11}	$4.883\ 325\ 436\ 4 \times 10^{-19}$			
a_{12}	$1.080\ 347\ 468\ 3 \times 10^{-21}$			
a_{13}	$1.394\ 929\ 102\ 6 \times 10^{-24}$			
a_{14}	$7.979\ 589\ 315\ 6 \times 10^{-28}$			

Table 3.6. Temperature as a function of emf for common thermocouples. The reference temperature is at 0°C. The power series expansion is of the form $T = \sum_1^n b_n V^n$, where V is the emf in microvolts and T is the temperature in degrees Celsius. The tabulated coefficients are for a quartic approximation over the indicated range; higher-degree equations are available for larger temperature ranges (Powell et al., 1974).

	T	K	J	E
Temperature range (°C):	0 to 400	−20 to 500	−20 to 500	−20 to 500
error (°C):	−0.15 to 0.17	−1.2 to 0.6	−0.07 to 0.06	−0.18 to 0.12
b_1	$2.566\ 129\ 7 \times 10^{-2}$	$2.436\ 385\ 1 \times 10^{-2}$	$1.974\ 505\ 6 \times 10^{-2}$	$1.697\ 028\ 7 \times 10^{-2}$
b_2	$-6.195\ 486\ 9 \times 10^{-7}$	$5.620\ 693\ 1 \times 10^{-8}$	$-1.809\ 425\ 6 \times 10^{-7}$	$-2.083\ 060\ 3 \times 10^{-7}$
b_3	$2.218\ 164\ 4 \times 10^{-11}$	$-3.882\ 562\ 0 \times 10^{-12}$	$7.877\ 791\ 9 \times 10^{-12}$	$4.651\ 271\ 7 \times 10^{-12}$
b_4	$-3.550\ 090\ 0 \times 10^{-16}$	$3.912\ 020\ 8 \times 10^{-17}$	$-1.189\ 722\ 2 \times 10^{-16}$	$-4.180\ 578\ 5 \times 10^{-17}$

Table 3.7. Thermal emf in microvolts for copper–constantan (type *T*) thermocouple material at various temperatures in degrees Celsius with reference temperature at 0°C (Roeser and Louberger, 1958).

°C	+0	+1	+2	+3	+4	+5	+6	+7	+8	+9	μV °C^{-1}
−20	−757	−720	−683	−646	−608	−571	−534	−496	−458	−421	37.4
−10	−383	−345	−307	−269	−231	−193	−154	−116	−77	−39	38.3
0	0	39	78	116	156	195	234	273	312	352	39.1
10	391	430	470	510	549	589	629	669	709	749	39.9
20	789	829	870	910	951	992	1 032	1 073	1 114	1 155	40.7
30	1 196	1 237	1 278	1 320	1 361	1 403	1 444	1 486	1 528	1 570	41.5
40	1 611	1 653	1 696	1 738	1 780	1 822	1 865	1 907	1 950	1 992	42.4
50	2 035	2 078	2 121	2 164	2 207	2 250	2 294	2 337	2 380	2 424	43.2
60	2 468	2 511	2 555	2 599	2 643	2 687	2 731	2 775	2 819	2 864	44.1
70	2 908	2 953	2 997	3 042	3 087	3 132	3 176	3 222	3 267	3 318	44.9
80	3 357	3 402	3 448	3 493	3 539	3 584	3 630	3 676	3 722	3 768	45.7
90	3 814	3 860	3 906	3 952	3 998	4 045	4 091	4 138	4 184	4 231	46.4

Table 3.8. Thermal emf microvolts for chromel–alumel (type K) thermocouple material at various temperatures in degrees Celsius with reference temperature at 0°C (Roeser and Louberger, 1958).

°C	+0	+10	+20	+30	+40	+50	+60	+70	+80	+90	$\mu V\,°C^{-1}$
−100	−3 553	−3 242	−2 920	−2 586	−2 243	−1 889	−1 527	−1 156	−777	−391	35.5
0	0	397	798	1 203	1 611	2 022	2 436	2 850	3 266	3 681	41.0
100	4 095	4 508	4 919	5 327	5 734	6 137	6 539	6 939	7 338	7 737	40.4
200	8 137	8 537	8 938	9 341	9 745	10 152	10 560	10 969	11 381	11 793	40.7
300	12 207	12 623	13 039	13 456	13 874	14 292	14 712	15 132	15 552	15 974	41.9
400	16 395	16 818	17 241	17 664	18 088	18 513	18 938	19 363	19 788	20 214	42.4
500	20 640	21 066	21 493	21 919	22 346	22 772	23 198	23 624	24 050	24 476	42.6
600	24 902	25 327	25 751	26 176	26 599	27 022	27 445	27 867	28 288	28 709	42.3
700	29 128	29 547	29 965	30 382	30 799	31 214	31 629	32 042	32 455	32 866	41.5
800	33 277	33 686	34 095	34 502	34 908	35 314	35 713	36 121	36 524	36 925	40.5
900	37 325	37 723	38 122	38 519	38 915	39 310	39 703	40 096	40 488	40 879	39.4
1 000	41 269	41 658	42 045	42 432	42 818	43 202	43 586	43 968	44 349	44 729	38.4
1 100	45 108	45 486	45 863	46 238	46 612	46 985	47 356	47 726	48 095	48 462	37.2
1 200	48 828	49 192	49 555	49 916	50 276	50 634	50 990	51 344	51 697	52 049	35.7
1 300	52 398	52 746	53 093	53 438	53 782	54 125	54 466	54 807	55 147	55 486	34.3

Table 3.9. Thermal emf in microvolts for iron–constantan (type J) thermocouple material at various temperatures in degrees Celsius with reference temperature at 0°C (Brown and Louberger, 1958).

°C	+0	+10	+20	+30	+40	+50	+60	+70	+80	+90	μV °C⁻¹
-100	-4 632	-4 215	-3 785	-3 344	-2 892	-2 431	-1 960	-1 481	-995	-501	46.3
0	0	507	1 019	1 536	2 058	2 585	3 116	3 649	4 186	4 725	52.7
100	5 268	5 812	6 359	6 907	7 457	8 008	8 560	9 113	9 667	10 221	55.1
200	10 776	11 332	11 887	12 442	12 998	13 553	14 108	14 663	15 217	15 771	55.5
300	16 325	16 878	17 432	17 984	18 535	19 088	19 640	20 192	20 743	21 294	55.2
400	21 846	22 397	22 949	23 501	24 054	24 607	25 161	25 716	26 272	26 830	55.4
500	27 388	27 949	28 511	29 075	29 642	30 210	30 782	31 356	31 933	32 513	57.1
600	33 096	33 683	34 273	34 867	35 464	36 066	36 671	37 280	37 893	38 510	60.3
700	39 130	39 754	40 382	41 013	41 647	42 283	42 922	43 563	44 206	44 852	63.7

tables. The limits of error for standard commercial type-J thermocouples are $\pm2.2°C$ between $0°–277°C$; the limits on premium wire are $\pm1.1°C$. The Seebeck coefficients near ambient temperatures range from $50.4\ \mu V°C^{-1}$ at $0°C$ to $54.3\ \mu V°C^{-1}$ at $100°C$. The seven-degree equation defining emf with respect to temperature can be obtained from National Bureau of Standards (1974); tabulated emf values at $10°C$ increments are summarized in Table 3.9. The coefficients of a fourth-degree equation used to estimate temperature ($°C$) from emf (μV) are tabulated in Table 3.6.

Chromel–constantan (type E) thermocouples have a large Seebeck coefficient that ranges from $58.7\ \mu V°C^{-1}$ at $0°C$ to $67.5\ \mu V°C^{-1}$ at $100°C$. The $E+$ material is the same as $K+$, and the E-material is the same as $T-$. It is evident in Fig. 3.7 that this combination would give the largest Seebeck coefficient of the materials illustrated. The use of this combination is of relatively recent origin. It is recommended for use in the range $-250°$ to $871°C$. The materials have relatively low thermal conductivity, good resistance to corrosion, and reasonably good homogeneity. The limits of error between $0°–316°C$ are $\pm1.7°C$ for standard wire and $\pm1.25°C$ for premium wire. The coefficients of a fourth-degree equation that can be used to estimate temperature ($°C$) from emf (μV) are tabulated in Table 3.6.

3.3.1.3   Elementary Thermocouple Circuits

Thermocouples are often used as a single loop with one reference and one measuring junction. However, thermocouples can be connected in series or in parallel to gain certain advantages. A number of similar thermocouples may be connected in series with all of their measuring junctions at T_2 and the reference junctions at T_1. Such a series is called a thermopile, and n

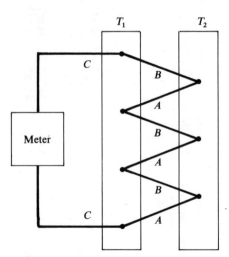

Figure 3.9. A multijunction thermopile. An example of a 3-junction thermopile made from materials A and B; C is leadwire.

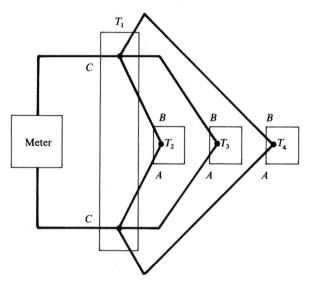

Figure 3.10. Parallel thermopile. An example of 3 thermocouples in parallel made from materials A and B; C is lead material.

thermocouples in series generates an emf n times as great as a single thermocouple (Fig. 3.9).

If n thermocouples are connected in parallel, the emf generated is the same as for single thermocouples. If all the thermocouples are of equal resistance and their measuring junctions are at various temperatures, T_2, T_3, etc., the thermal emf generated will correspond to the mean of the temperatures of the individual junctions (Fig. 3.10). It is not necessary to adjust the thermocouple resistance when measuring these average temperatures. Instead, swamping resistors may be used. For example, if the thermocouples range in resistance from 8 to 12 Ω, a 200-Ω $\pm$ 1 % resistor can be connected in series with each, and the error in emf introduced by inequality in the thermocouple resistance may be an insignificant fraction of the total.

3.3.1.4 Thermocouple Reference Junctions

When accurate thermocouple measurements are required, it is common practice to maintain the thermocouple reference junction at some constant temperature so that only copper leads are connected to the emf readout instrument. This procedure avoids the generation of thermal emf at the terminals of the readout instruments. Changes in reference junction temperature influence the output signal and practical instruments must be provided with a means to cancel this potential source of error.

The constant temperature is usually the melting point of ice (0°C). However, constant temperature ovens that are maintained above ambient are frequently used. When ovens are used, the emf of a thermocouple at

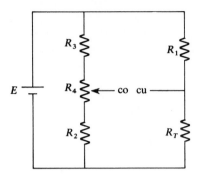

Figure 3.11. Thermocouple cold junction compensator. For copper–constantan: $R_1 = R_3 = 2\,000\ \Omega$; R_2 and $R_4 = 50\ \Omega$; R_4 is a 20-turn potentiometer; $R_T = 16.12\ \Omega$ at 20°C (approximately 7.47 m of 38-gauge copper wire); $E = 1.35\text{-V}$ mercury battery; co = constantan connection; and cu = copper connection. For chromel–constantan: $R_1 = R_3 = 2\,000\ \Omega$; R_2 and $R_4 = 50\ \Omega$; R_4 is a 20-turn potentiometer; $R_T = 24.4\ \Omega$ at 20°C (approximately 11.31 m of 38-gauge copper wire); co and cu are copper connections; the chromel–copper and the constantan–copper connections must be at the same temperature; and $E = 1.35\text{-V}$ mercury battery.

that temperature (e.g., 1 611 μV at 40°C, Table 3.7) must be added to the measured emf (e.g., $-822\ \mu$V at 20°C) to obtain the unknown temperature using tables referenced at 0°C (e.g., $1\,611 - 822 = 789\ \mu\text{V} \simeq 20°\text{C}$).

Electronic reference junctions are now available and may be used in place of ice baths or ovens. The electronic junctions operate from batteries or ac and are available in sizes as small as a piece of chalk. Basically, they contain

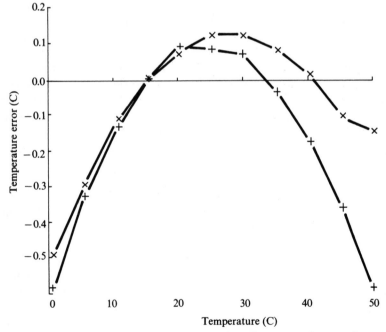

Figure 3.12. Temperature error (bridge output minus thermocouple output) expressed in °C of the compensators shown in Figure 3.11, where + is copper–constantan and × is chromel–constantan.

a bridge (Fig. 3.11) that generates a signal equal to and opposite in polarity to the thermocouple reference junction encapsulated with the active leg of the bridge. Over the temperature range 10°–40°C, the error is ±0.1°C for chromel–constantan thermocouple and ±0.2°C for copper–constantan thermocouple (Fig. 3.12).

3.3.1.5 Constructing Thermocouple Junctions

Thermocouple junctions may be made with mechanical fasteners, by welding, with silver solder, or with soft solder. If properly done, the emf is the same from junctions formed by any of the methods.

Soft soldered junctions are most common in the environmental range. Rosin core flux electronic solder may be used for the environmental temperature range. Some alloys are difficult to solder; consequently, an additional cleaner or scraping is required. If a cleaner is used, the wires should be washed with trichlorethylene to prevent further corrosion.

Junctions bonded by silver solder are mechanically stronger than those bonded with soft solder. Silver solder bonding consists of brazing at low temperatures, around 550°–650°C. Brazing material is usually for specific metals or alloys. However, some general purpose brazes are available for the types of metals and alloys used for thermocouples.

Materials to be brazed must be cleaned. A flux is used that is liquid when heated to the brazing temperature. The brazing rod is then applied to the heated area; it will flow to heated areas that are clean and covered with flux.

Welded thermocouple junctions are the strongest and are necessary for measurement of fire temperature. Methods of welding consist of mechanical contact, oxy-gas, percussion, resistance, and arc welding. Commercial welding equipment is available. However, four types of welding are described that can be performed in most laboratories. The choice of method depends on the wire size and the type of junction desired.

Percussion welding is used for fine wires (<0.2 mm). The technique is described by LeMay (1958). An arc is established by the discharge of a capacitor as the two wires are moved close to each other. This arc brings each surface up to welding temperature before they contact. When the parts meet the arc is extinguished, and the propelling force and momentum provide the follow-up necessary to join the parts. Wires as small as 0.2 mm were aligned for butt welding inside a small glass tube. Percussion welding is preferred over resistance welding because less energy is required and there is less chance of metallurgical changes that may weaken the material.

Resistance welding consists of passing a current through the wires to be welded after they have been aligned and pressure has been applied to the junction. This technique is recommended for butt welds because only a minute portion of the wire is affected (Hickson, 1940). Stover (1960) described the use of a micromanipulator and a capacitor discharge network for welding wires as small as 0.02 mm.

Arc welding is used for lap joining the wires. One method requires a small amount of mercury covered with a thin layer of oil in a metal container and should be performed in a hood. About 1 cm of insulation is stripped from the ends of the thermocouple wires. The wires are firmly twisted together and an alligator clip is used to grasp the bare wire between the twisted portion and the insulation. The clip is connected to one lead from a 24-V transformer, similar to those used in home furnaces. The other transformer lead is clipped to the metal container. The twisted portion of the wire is used to touch the mercury. A flash and a welded junction will result.

An arc welder may be used to weld larger wires. A carbon rod or pencil lead is used as the electrode. The twisted wire is stroked with the electrode to form the junction. Lopushinsky (1971) described a jig for arc welding 0.025-mm chromel and constantan wires.

3.3.1.6   Thermocouple Precautions

As a general rule, double insulated wire should always be used because of the possibility of a scrape in the insulation and ground potential pickup. Shielding adds mechanical strength to wires and should be used even though it is not necessary in many cases.

Ground potential pickup may completely swamp the true signal when the soil is wet and be unnoticed when it is relatively dry. To check on insulation leaks, one should measure the resistance to ground of the sensor and lead wire. A resistance to ground of 500 kΩ or less can result in signal errors.

Some researchers use common wires and switch only one side of the circuit when measuring multiple temperatures. This tempting practice reduces cost but should be avoided because if an insulation leak develops in any of the commoned circuits, the whole group is subject to error.

When passing thermocouple leads through panels where it is possible for temperature gradients to exist, one should always use thermocouple connectors. Connectors of other materials may introduce four additional thermocouple junctions (see Sect. 3.3.1.1).

Insulation may be removed from wire with a hot soldering iron. If a wire stripper is used, care should be taken to avoid nicking the wire which may cause breakage. Solid conductors should be avoided except for permanent installations. Stranded thermocouple duplex wire is available, individual strands of which can be used for junctions. The parallel configuration of duplex leads is not as efficient in rejecting electromagnetic radiation as is a twisted pair.

3.3.2   Resistance Elements

The electrical resistance of most materials varies with temperature. This characteristic is utilized to measure temperatures with resistance thermometers made from a variety of materials. Given a known resistance–temperature relationship and appropriate recording equipment, temperatures can be determined to a high degree of precision. Choice of suitable materials depends on the size and linearity of the resistance coefficient (dR/dT) and such other factors as the absolute resistance, workability, stability, and thermal emf against copper.

Many different materials have been used in resistance thermometers. Most metals are distinguished by positive resistance coefficients, so the resistance increases with temperature. Several types of semiconductor elements (thermistors, transistors, diodes) have been used successfully for temperature measurements. They exhibit a negative resistance coefficient, and the resistance decreases with temperature. The semiconductor materials have been gaining popularity in recent years.

3.3.2.1 Resistance Thermometers

Resistance thermometers have been used for many years for precision temperature measurements. The relation between resistance and temperature is described for most metals by the following relationship:

$$R_T = R_0[1 + \alpha(T - T_0) + \beta(T - T_0)^2] \tag{3.9}$$

where R_0 is the resistance at T_0 usually $0°C$, R_T is the resistance at T, and α and β are first and second order temperature coefficients. The linear term alone may be satisfactory approximation in some situations. The first- (α) and second- (β) order temperature coefficients for some commonly used metals are given in Table 3.10.

Resistance thermometers are commonly constructed by winding fine wire ($5-100$-μm diameter) onto a form until the length is sufficient to achieve the desired resistance. The resistance of these fine wire elements is affected by strain, so the winding forms must have a small coefficient of expansion. Glass and mica are commonly used. The elements are also annealed after fabrication to relieve winding strains that may have developed. Construction of resistance elements is not technically difficult, but they are seldom custom made as many different types are available commercially at rather low prices.

Table 3.10. First- and second-order temperature coefficients, resistance, emf, and winding size for some common metals (Tanner, 1963).

| Metal | Temperature coefficients | | Resistance at 293 K[a] | Emf (mV) | Winding size (μm)[b] |
	First (α)	Second (β)			
Platinum	3.9×10^{-3}	-0.55×10^{-6}	9.8×10^{-6}	-7.6	$25-50$
Nickel	$4-5 \times 10^{-3}$	7.5×10^{-6}	$6-10 \times 10^{-6}$	-22.0	$25-50$
70% Ni, 30% Fe[c]	4.6×10^{-3}	3.5×10^{-6}	22×10^{-6}	-39.0	$25-50$
Tungsten	4.5×10^{-3}	0.5×10^{-6}	5.5×10^{-6}	$+3.6$	$5-10$
Copper	4.0×10^{-3}	0	1.7×10^{-6}	0	$50-100$

[a] In Ω cm^{-3}.
[b] Minimum convenient winding size.
[c] Balco or Hytempo are trade names.

Table 3.11. Resistance of a 100-Ω platinum resistance element according to the IPTS-68.

°C	0	1	2	3	4	5	6	7	8	9	Ω °C^{-1}
−20	92.0	92.4	92.8	93.2	93.6	94.0	94.4	94.8	95.2	95.6	0.400 3
−10	96.0	96.4	96.8	97.2	97.6	98.0	98.4	98.8	99.2	99.6	0.399 1
0	100.0	100.4	100.8	101.2	101.6	102.0	102.4	102.8	103.2	103.6	0.397 9
10	104.0	104.4	104.8	105.2	105.6	106.0	106.4	106.8	107.2	107.5	0.396 8
20	107.9	108.3	108.7	109.1	109.5	109.9	110.3	110.7	111.1	111.5	0.395 6
30	111.9	112.3	112.7	113.1	113.5	113.9	114.3	114.7	115.1	115.5	0.394 4
40	115.8	116.2	116.6	117.0	117.4	117.8	118.2	118.6	119.0	119.4	0.393 2
50	119.8	120.2	120.6	121.0	121.3	121.7	122.1	122.5	122.9	123.3	0.392 1
60	123.7	124.1	124.5	124.9	125.3	125.7	126.0	126.4	126.8	127.2	0.390 9
70	127.6	128.0	128.4	128.8	129.2	129.6	129.9	130.3	130.7	131.1	0.389 7
80	131.5	131.9	132.3	132.7	133.1	133.4	133.8	134.2	134.6	135.0	0.388 5
90	135.4	135.8	136.2	136.5	136.9	137.3	137.7	138.1	138.5	138.9	0.387 4
100	139.3	139.6	140.0	140.4	140.8	141.2	141.6	142.0	142.3	142.7	0.386 2

Platinum is preferred for resistance thermometers because it is available in very pure form, is stable, and has a relatively low thermal emf and second-order coefficient. However, the absolute resistance value is also low. Platinum elements of 100-Ω nominal resistance are available encapsulated in glass with dimensions as small as 1.4 mm diameter and 14 mm long. A 100-Ω platinum resistance element is shown in Fig. 3.3(b). The temperature relationship for platinum was given by Callender (1887) as

$$T = \frac{1}{a}\left(\frac{R_T}{R_0} - 1\right) + b\left(\frac{T}{100} - 1\right)\frac{T}{100}, \tag{3.10}$$

where R_T and R_0 are resistances at temperatures T and 0°C. This relation was modified by the International Practical Temperature Scale of 1968 (IPTS-68) to

$$T_{68} = \frac{1}{a}\left(\frac{R_T}{R_0} - 1\right) + b\left(\frac{T}{100} - 1\right)\left(\frac{T}{100}\right)$$
$$+ 0.045\left(\frac{T}{100}\right)\left(\frac{T}{100} - 1\right)\left(\frac{T}{419.58} - 1\right)\left(\frac{T}{630.74} - 1\right),$$

where $a = 3.925\,966\,8 \times 10^{-3}°C^{-1}$ and $b = 1.496\,334°C$ (Riddle et al., 1973). Nominal resistances for a 100-Ω platinum resistance thermometer are presented in Table 3.11.

Recently, nickel film resistors that are less expensive than platinum have become available. They are nearly linear from $-20°$ to 60°C, with maximum deviations of 0.7%. The nominal resistance of 1 000 Ω has a temperature coefficient of 0.005 9Ω °C^{-1}.

With reasonable care, resistance thermometers can be used to determine temperature to 0.1°C. Self-heating is not a serious problem if the current is kept below 2 mA. The design of a 100-Ω platinum element will allow about 4 mW to dissipate for each degree of temperature rise. Self-heating errors should thus remain below 0.1°C for a current of 2 mA in the element.

3.3.2.2   Thermistors

Thermistors are semiconductors of ceramic materials made from sintering mixtures of oxides of manganese, nickel, cobalt, copper, iron, and uranium. Thermistors have very large negative temperature coefficients; each temperature increase of one degree Celsius will decrease the resistance five percent. Thermistor resistance change with temperature is much larger than for resistance thermometers, however, the change with temperature is quite nonlinear. Thermistor characteristics have been known for many years, but only in the last 20 years have elements been produced with stable and reproducible characteristics.

Thermistors are available in many sizes and in shapes such as beads, discs, washers, and rods. Commercially available beads range in diameter from 0.015 to 2.5 mm (Fig. 3.3(c)]. Thermistors of small size and low specific heat draw practically no energy from the object to be measured, so they are well suited for localized temperature measurements.

The resistance of thermistors is a function of the absolute temperature, expressed as

$$R(T) = R(T_0) \exp B\left(\frac{1}{T} - \frac{1}{T_0}\right), \qquad (3.11)$$

where $R(T)$ is the resistance at absolute temperature T, $R(T_0)$ is the resistance at absolute temperature T_0 (usually taken at $T_0 \cong 298.16°K$), and B is a constant around 4000 that varies somewhat with the thermistor material. Typical resistances for a 3000-Ω thermistor at 25°C are given in Table 3.9.

The self-heating errors in a thermistor can be quite large because of the large temperature coefficient. For example, the relationship between current applied and resistance of a Fenwal 2000-Ω glass bead thermistor is illustrated in Fig. 3.13. Starting at the lower left, an increase in current applied causes no change in the thermistor resistance until point A is reached. At this point, enough heat is generated to lower the resistance of the thermistor and cause more current to flow, which lowers the resistance more. Eventually, a point is reached at which the power dissipated equals the rate at which thermal energy can be lost from the heated thermistor. The safe power dissipation

Table 3.12. Resistance of a 3000-Ω Yellow Springs Instrument thermistor 44 005.

Temperature (°C)	Resistance (Ω)
−20	29 130
−15	21 890
−10	16 600
−5	12 700
0	9 796
5	7 618
10	5 971
15	4 714
20	3 748
25	3 000
30	2 417
35	1 959
40	1 598
45	1 310
50	1 081

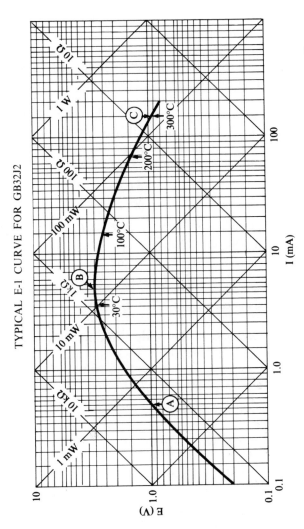

Figure 3.13. Resistance of a Fenwal 2 000-Ω glass bead thermistor in relation to voltage, *E*, and current, *i*, applied. Ⓐ Max current for no appreciable self heat. Ⓑ Peak voltage. Ⓒ Max safe continuous current items shown are thermistor (Temp.). (Courtesy of Fenwal Electronics, Inc.)

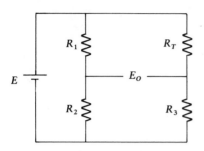

Figure 3.14. Wheatstone bridge. R_T is the temperature sensitive resistance element; R_1, R_2, and R_3 are fixed resistors; E is voltage applied and E_0 is voltage out.

of this thermistor appears to be about 0.5 mW. One milliwatt is a good general estimate of the energy that can be dissipated from a bead in moving air.

The data in Table 3.12 can be used to design a thermistor bridge with a nearly linear scale from $-10°$ to $+40°C$. The schematic for the bridge is shown in Fig. 3.14, where R_T represents the thermistor resistance.

First, consider the circuit at the two temperature extremes. The voltage drop across the thermistor at $-10°C$, using Eq. (2.44), is

$$E_1 = \frac{aE}{a + R_3} = \frac{16\,600}{16\,600 + R_3}, \tag{3.12}$$

where $R_T = a$, the resistance at $-10°C$ and $R_2 = R_3$. The voltage drop across the thermistor at 40°C is

$$E_2 = \frac{bE}{b + R_3} = \frac{1\,598}{1\,598 + R_3} \tag{3.13}$$

where $R_T = b$, the resistance at 40°C. The thermistor value at midrange (15°C) is $c = 4\,714\ \Omega$.

Next, evaluate the voltages in the bridge. The maximum difference in voltage is

$$\Delta E_o = E_2 - E_1 = \frac{aE}{a + R_3} - \frac{bE}{b + R_3}, \tag{3.14}$$

and the voltage over half the range is

$$\tfrac{1}{2}\,\Delta E_o = \frac{aE}{a + R_3} - \frac{cE}{a + R_3}. \tag{3.15}$$

Now, the unknown resistor R_3 is determined by combining Eqs. (3.14) and (3.15);

$$R_3 = \frac{bc + ac - 2ab}{a - 2c + b} = 3\,732\ \Omega. \tag{3.16}$$

The applied voltage, E, is given by

$$E = \Delta E_o \frac{(R_3 + a)(R_3 + b)}{a(R_3 + b) - b(R_3 + a)} = 1.94\ \Delta E_o. \tag{3.17}$$

The dissipation constant of the thermistor in still air is $1\ mW°C^{-1}$. For a self-heating error of 0.1°C, the applied power should not exceed 0.1 mW.

The maximum power will be dissipated in the thermistor when the thermistor resistance is the same as that of a resistor in series with it, that is, 3 732 Ω. Therefore, the applied thermistor voltage, $E_T^2 = PR = 0.000\,1(3\,732) = (0.373\,\text{V})^2$ or $E_T = 0.611\,\text{V}$, and the voltage applied to this bridge is 1.222 V. A 1.35-V mercury battery would be an excellent power supply.

A maximum sensitivity of 13.9 mV°C^{-1} is achieved by measuring the potential difference across the bridge. However, the impedance would be around 4 200 Ω. The impedance may be reduced by measuring the voltage drop across a portion of a shunt resistor located in place of E_o in Fig. 3.14. If 100 kΩ were used for the shunt, the current flow through the shunt would be $i = \pm 0.348/100\,000 = \pm 3.48\,\mu$A. Measurement of the voltage drop across 725 Ω of the shunt would result in a sensitivity of 100 μV°C^{-1} or ± 2.5 mV full scale.

For maximum linearity, R_1 should equal R_T at midrange, e.g., 4 714 Ω at 15°C. However, the output of the bridge would be zero at 15°C instead of at 0°C. To obtain a zero bridge output at 0°C, R_1 would have to be 9 796 Ω. Linearity of output versus temperature is shown in Fig. 3.15. The bridge is designed to have perfect agreement at three temperatures: $-10°$, 15°, and 40°C. The greatest deviation from linearity of 1.1°C occurs at 0°C.

Parallel thermistors with series resistors can be used to improve linearity. An example of a thermal-linear network is shown in Fig. 3.16. According to Trolander (1972), this network yields a linear output (± 0.08°C from $-50°$ to 50°C).

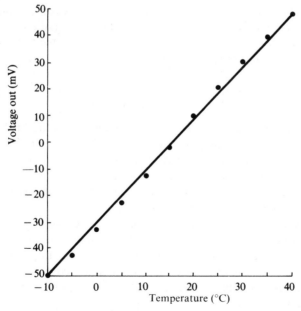

Figure 3.15. Voltage output of a thermistor temperature bridge versus temperature. Dots are experimental values.

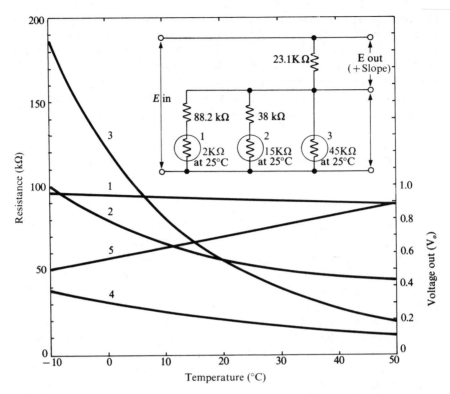

Figure 3.16. Thermal-linear network (Trolander, 1972). Resistance of individual components are plotted versus temperature where 1 is 88.2 kΩ plus a 2-kΩ thermistor; 2 is 38 kΩ plus a 15-kΩ thermistor; 3 is a 45-kΩ thermistor; 4 is the parallel resistance of 1, 2, and 3, and 5 is voltage out when voltage in is 1.35 V.

3.3.2.3 Minimizing Bridge Errors

The unknown resistance, R_T, should be connected to the bridge with low-resistance leads. Since the lead resistance adds to the element resistance, the calibration may be in error. This can become a problem with long, small diameter leads. Also, the resistance in the lead wires will vary with temperature. The lead resistance changes may introduce spurious readings, even though the temperature coefficient of resistance of copper is quite low.

The effects of lead resistances can be reduced with the simple 3-wire bridge illustrated in Fig. 3.17(a), and can be excluded completely with special 4-wire circuits [Fig. 3.17(b)]. Sometimes, it may be simpler to calibrate a 2-wire bridge with leads in place than to calculate corrections. Corrections are generally most applicable when the lead resistance is 10% or less of the sensor resistance.

The current through the bridge may also induce errors by heating the resistors. This is especially true for the resistance element being used to

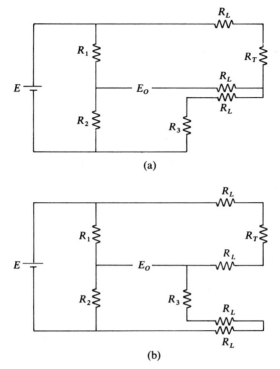

Figure 3.17(a) and **3.17(b).** Multiwire Wheatstone bridge configurations. (a) Three-wire Wheatstone bridge where R_L is the leadwire resistance. (b) Four-wire Wheatstone bridge.

measure temperature since it necessarily has a larger temperature coefficient than do the other bridge resistors.

The power, P, dissipated in a resistor is

$$P = VA = \frac{V^2}{R}. \tag{3.18}$$

In a bridge configuration, the maximum power will be generated when the resistance element equals the value of the bridge resistor in series with it, e.g., $R_T = R_3$ in Fig. 3.14. It is, therefore, quite simple to calculate the maximum energy dissipated in the circuit but somewhat more difficult to estimate the temperature error for a given rate of power dissipation.

The temperature increase due to self-heating depends primarily on the rate at which the dissipated energy can be transferred to the environment. Thus, the temperature increase will be less for elements in water than for those in air and less for those in turbulent flow than for those in still air.

The temperature increase can be estimated for a given level of heat dissipation from a basic heat transfer analysis of the specific conditions for that probe. The true temperature, T_t, can also be estimated empirically by observing the temperature change associated with several levels of current flow,

$$T_t = T_1 + \frac{(T_2 - T_1)i_1{}^2}{(i_1{}^2 - i_2{}^2)}, \qquad (3.19)$$

where T_1 and T_2 are the temperatures observed with currents of i_1 and i_2.

3.3.2.4 Diodes

Diodes and transistors have been used for temperature sensors (Barton, 1962; McNamara, 1962; Dimick and Trezek, 1963). The forward conducting voltage drop across a diode junction varies with temperature and current. Sargeant (1965) found the voltage drop across a forward biased germanium (IN2326) diode to be approximately -2.3 mV°C^{-1} with a 1 mA current. The sensitivity was linear from 0° to 40°C but tailed off above 60°C.

Hinshaw and Fritschen (1970) studied silicon diodes [Fig. 3.3(d)] and found the temperature sensitivity to be constant over a much larger temperature range. The temperature sensitivity is dependent on the current as shown in Fig. 3.18. A current of 0.5 mA was selected to (1) keep the operating resistance of the diode less than 1 000 Ω, (2) maintain a large temperature sensitivity not greatly dependent on current stability, and (3) have low self-heating.

Desirable characteristics of diodes to be used as temperature sensors include high conductivity, low leakage, and glass encapsulation. Examples are the Fairchild FD-300 and the General Electric SSD-708. When FD-300 diodes were tested in the circuit shown in Fig. 3.19, The forward voltage drop at 0°C was 670 mV and the sensitivity was -1.97 mV °C^{-1}. Under field conditions, the voltage drop along the lead wires may be significant

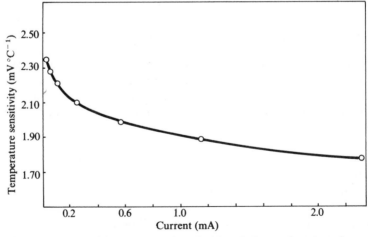

Figure 3.18. Temperature sensitivity of a FD-300 diode as a function of current.

Figure 3.19. Diode thermometer circuit. E is 22 V dc and R is 42 kΩ.

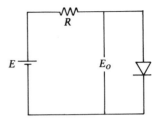

and should be subtracted from the total voltage drop. Circuits similar to the 3- and 4-wire bridges may be used to minimize this effect.

3.3.3 Comparison of Electrical Temperature Sensors

The final choice of a temperature sensing element depends on many factors, not the least of which is the measuring device or recording equipment available. Some factors that should be considered before making the final choice are listed in Table 3.13.

Thermocouples constructed from a single spool of premium wire will have the same characteristics. If one is calibrated, the same calibration will apply to all from the same spool and the standard thermocouple emf tables will usually apply.

Thermocouples are low impedance devices that can be used with most amplifiers and recorders. The main disadvantage of the thermocouple is the low-level signal. A 10-μV recording resolution equal to 0.25°C is adequate for many uses.

Although resistance elements can be arranged to have larger outputs than thermocouples, each one must be calibrated individually. The resistance of the lead wire must be considered although errors associated with changes in wire resistance can easily be eliminated.

Thermistors have larger resistance coefficients than resistance elements, and can yield a larger sensitivity. They have greater nonlinearity and special circuits are needed if a linear output is desired. Thermistors also have to be calibrated individually when used for temperature measurements.

Silicon diodes yield the most linear response of all of the temperature sensors when used in the primary state. Furthermore, diodes from a given manufacturing lot have about the same sensitivity. Diodes used as shown in Fig. 3.19 do not require elaborate calibration for climatological use. The voltage drop across the diode should be evaluated near the middle of the expected temperature range so that slight deviations in sensitivity will not cause large temperature errors to either side of the midpoint. For temperature gradient measurements, diodes should be individually calibrated at two points at least, e.g., 0° and 50°C.

Resistance elements, thermistors, and diodes require a precision power supply, as variations in applied voltage or current create temperature errors. The cost of precision power supplies has dropped considerably with the advent of integrated circuit components.

Table 3.13. Relative comparison of electrical temperature sensing elements.

Characteristics	Thermocouple	Resistance	Thermistors	Diodes
Physical size	small	large	small	medium
Wire size	small to large	very small	very small	medium
Signal	0.04 mV °C^{-1}	up to 100 mV °C^{-1} with proper bridge	250–500 mV °C^{-1} with proper bridge	2 mV °C^{-1}
Stability	excellent	fair to excellent	good	good
Linearity	slightly nonlinear	slightly nonlinear	very nonlinear	linear
Interchangeability	excellent	good	poor	good
Sensing at a point	excellent	poor	excellent	good
Power source	none	required	required	required
Series arrangement	easy	easy	no	easy
Parallel arrangement	required	need to match sensors	need to match sensors	need to match sensors
Temperature reference	required	no	no	no
Expense	least	high	medium	low

Based upon many factors and on a great deal of practical experience, we feel that thermocouples are still the simplest and most reliable temperature sensors.

3.4 Air Temperature

The selection of a specific sensor and the design of the measuring system depend on the purpose of the measurement. The accuracy requirements are important, as is the location of the measurement, e.g., in the vegetative canopy, near the surface of the soil or leaves, or in the atmosphere well above the canopy. Accuracy requirements are often unrealistically specified. For example, Tanner (1963) relates his experience with biologists who initially insisted on an accuracy of $\pm 0.01°C$, but who rapidly degraded their requirements to $\pm 2°C$ when they learned of the expense, labor, and sampling problems associated with the more precise measurements. All measurement users need a fuller comprehension of biological requirements as well as an understanding of the physics of measurement.

3.4.1 Sources of Error

The choice of a recording or readout mechanism may affect the accuracy as much as the choice of the sensor. We shall give special consideration to the sensor design, however, to insure that temperature measurements are representative. Four types of errors must be considered in temperature sensor design: velocity, conduction, transient, and radiation.

3.4.1.1 Velocity Errors

The velocity error, E_v, which is the difference between the true temperature, T_t, and the sensor temperature, T, is expressed as

$$E_v = \frac{(1 - r)U^2}{2gJc_p},$$

(3.20)

where r is the recovery factor, U is the velocity, g is the gravitational constant, J is the mechanical equivalent of thermal energy (0.238 9 cal/J), and c_p is the specific heat of the fluid at constant pressure.

The recovery factor, r, is defined as the difference between the equilibrium temperature sensed by a stationary ideal-geometry, adiabatic probe, and the static temperature (actual temperature of the gas at all times) divided by the dynamic temperature (thermal equivalent of the directed kinetic energy of the gas continuum). It is related to the Prandtl number, Pr. For turbulent flow it is nearly $Pr^{1/3}$ and for laminar flow it is nearly $Pr^{1/2}$.

When a thermometer is exposed in a high velocity stream of air, friction and adiabatic compression increase the temperature. This increase is proportional to the square of the velocity in the stream and is influenced by the thermometer's shape and size. The error should be less than 0.02°C in air at velocities of 10 m/s, rising to perhaps 1°C at velocities of 100 m/s. The velocity error can generally be ignored in environmental measurements of air temperature.

3.4.1.2 Conduction Errors

Thermal energy will be conducted to or from the temperature sensor whenever the sensor is in contact with surfaces that differ from the true temperature. Commonly, the supports or lead wires of a temperature sensor provide a pathway for conduction of energy. Conduction problems may also appear when a sensor is inserted into another material. For example, how far should a thermocouple be immersed into water to ensure that the junction is at the temperature of the fluid and is not influenced by a differing air temperature?

Still another conduction error can occur when the tip of a thermocouple is pressed against a warm object whose temperature is to be measured (e.g., a tree needle). The thermocouple will conduct energy from the object. If this conduction rate is large with respect to the object's ability to supply thermal energy, the temperature at the point of contact will be diminished. Conduction error, E_c, can be estimated as the differences between true temperature T_t and the sensor temperature T,

$$E_c = \frac{T_t - T_m}{\cosh\, l(4h/d\lambda)^{1/2}}, \tag{3.21}$$

where l is the length of the wire between T_m and T, h is the convective heat transfer coefficient, d is the diameter of the wire, λ is the thermal conductivity, and T_m is the temperature of the environment or supports at a distance l from the sensor.

Conduction errors can be reduced by decreasing (a) the difference between true temperature and the mount temperature, (b) the wire diameter d, and (c) the thermal conductivity, or by increasing (d) the convective heat transfer coefficient, and (e) the junction length l. Of these many parameters, only the junction length is freely variable. Less latitude exists in selecting wire diameter and material with low conductivity. There is often little freedom for selection of either sensor supports or the size of convective heat transfer, h. Therefore, the designer must minimize errors by using small wires of low thermal conductivity and by maintaining a stable turbulent-aspiration flow. The other conduction errors can be reduced by fully immersing the sensor and by increasing the length of the lead wire between the supports and the exposed junction.

3.4.1.3   Transient Errors

Transient errors, E_t, result from the failure of a sensor to respond fully and rapidly to fluctuating temperatures. They are the difference between the true temperature T_t and the sensor temperature T,

$$E_t = \frac{\rho c \; d(dT)}{4h(dt)},\tag{3.22}$$

where ρ is the density of the air, c is the specific heat of the solid, d is the diameter of the solid, h is the convective heat transfer coefficient, and t is time.

Most environmental temperatures represent time averages. Except for eddy transport measurements and possibly a few special measurements, fast response temperature sensors are less desirable than temperatures averaged over a few minutes.

A thermometer value actually represents a period other than the instant of reading, with the lag related to the time constant of the thermometer. The time constant varies approximately inversely to the square root of the ventilation. As a convenient guideline, smoothed air temperatures can be obtained with sensors having a 20- to 60-s time constant ventilated at a rate of 4 m s^{-1}. Mean temperatures may be obtained by increasing the sensor's diameter and thermal capacity and by decreasing the convective heat transfer coefficient. The latter can be done by decreasing the aspiration rate. Covering the sensor with an insulating material will also introduce a lag into its response.

3.4.1.4   Radiation Errors

The radiation error, E_r, may be estimated by

$$E_r = \frac{K_r \sigma \varepsilon A_r (T_t^4 - T_w^4)}{h A_c},\tag{3.23}$$

where K_r is the view factor of the temperature sensor, σ is the Stefan–Boltzmann constant, ε is the emissivity, A_r is the radiational area, h is the convective heat transfer coefficient, A_c is the area available for convective heat transfer, T_t is the true temperature, and T_w is the temperature of the radiating surface.

The effect of radiation absorption on the temperature of bare wire at various wind speeds can be calculated from the steady state energy balance equation as follows:

$$\alpha\left(\frac{S\downarrow}{\pi} + 0.5D + 0.5K\uparrow\right) + \varepsilon(0.5L\downarrow + 0.5L\uparrow - \sigma T_s^4) + h(T_s - T_a) = 0.\tag{3.24}$$

The terms are defined as: α, absorption coefficient of short wave radiation; $S\downarrow$, direct beam radiation; D, diffuse radiation; $K\uparrow$, reflected solar radiation; ε, emissivity; L, longwave radiation down $\downarrow$ and up $\uparrow$; h, convective heat transfer coefficient; T_s, surface temperature; and T_a, air temperature. Substituting the assumption that

$$T_s^4 \cong -3T_a^4 + 4T_a^3 T_s \cong T_a^4 + 4T_a^3(\Delta T), \qquad (3.25)$$

and

$$h = 6.17 \times 10^{-3} \frac{U^{1/3}}{d^{2/3}}, \qquad (3.26)$$

where U is the wind speed and d is the diameter of the wire. Substitution of Eqs. (3.25) and (3.26) into Eq. (3.24) leads to

$$\Delta T = \frac{\alpha(S\downarrow/\pi + 0.5D + 0.5K\uparrow) + (0.5L\downarrow + 0.5L\uparrow - \sigma T_a^4)}{4\varepsilon\sigma T_a^3 + h}. \qquad (3.27)$$

Evaluation of Eq. (3.27) for a polished copper wire and a white painted wire on a hot sunny day over burned grass stubble and over transpiring grass leads to Figs. 3.20(a) and 3.20(b). The white paint was assumed to be 0.007 6 cm thick. The temperature excess (ΔT) is about the same for polished copper wire over either surface. However, the effect of longwave radiation absorption by white paint over the two surfaces is clearly demonstrated. White paint decreases the temperature excesses over cooler surfaces such as water or transpiring crops but increases the temperature excess of fine wire over hot surfaces such as burned stubble and dry soil.

3.4.1.5   Air Temperature Sensor Design

Air temperature sensor designs must be appropriate for the intended use. Sensors used in plant canopies, couvettes, or animal burrows should be of small size and have low ventilation requirements. In such small enclosures, out of direct sunlight, radiation errors will be small. Small polished wires have a high convective heat transfer relative to radiation absorption and may be used without additional radiation shielding. Natural ventilation often is desirable here because aspiration could alter the temperature regime of interest.

Radiation shielding becomes more important for the measurement of air temperature in the open. Very fine unshielded wires have been used satisfactorily. The design criteria depends somewhat on the choice of natural or forced ventilation. Concurrent measurements of humidity, for example, will usually require forced ventilation.

The two basic types of radiation shields are stacked flat plates and concentric tubes. The radiation error may be reduced by approximately $1/(1 + n)$

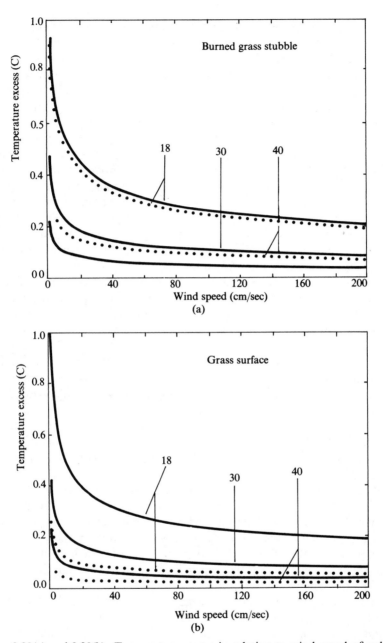

Figure 3.20(a) and **3.20(b)**. Temperature excess in relation to wind speed of polished copper wires of 18 (0.01 cm), 30 (0.05 cm), and 40 (0.007 6 cm) gauge (solid line) and of white painted wire (dotted line). The white paint is 0.007 6 cm thick. (a) Environmental conditions over burned stubble: $\alpha = 0.25$ and $\varepsilon = 0.15$ for polished copper wire while $\alpha = 0.20$ and $\varepsilon = 0.91$ for white paint; $T_a = 31.7°C$; $S\downarrow = 835$ W m^{-2}: $D = 127$ W m^{-2}; $K\uparrow = 50$ W m^{-2}; $L\downarrow = 372$ W m^{-2} and $L\uparrow = 628$ W m^{-2}. (b) The grass was assumed to be at air temperature and $L\uparrow$ was decreased to 493 W m^{-2}.

where n is the number of radiation shields (King, 1943). This shielding efficiency actually applies to conditions in a vacuum in which temperature distribution through successive layers of shielding is determined by radiation alone. Convective heat transfer in air augments the radiation exchange and makes the shields more effective. Radiation shields should be spaced to allow adequate air to flow between them.

Materials for constructing radiation shields are selected on the basis of their thermal properties. The top side of the upper shield is exposed to both short- and long-wave radiation, so it should have a low absorption for shortwave radiation and high emissivity for longwave radiation. Thus, the energy from absorbed solar radiation is freely reradiated from the surface as longwave radiation. Fuchs and Tanner (1965) examined radiative characteristics of various materials. They recommended, in the following order, aluminized Mylar, white paint, or a clear plastic coating on the upper surface of polished metal. These finishes were greatly superior to polished metal surfaces that are commonly used for shields.

The bottom of the outer shield is exposed to reflected shortwave and to longwave radiation. Since longwave radiation is more significant than shortwave radiation on the lower surface, the shield should have low shortwave absorptivity and longwave emissivity. Polished metals have these characteristics.

Since the inner surfaces of the outer shields, both surfaces of the inner shields, and the outer surface of the temperature sensor are exposed to longwave radiation, they should have low emissivity for longwave radiation to minimize the transfer of energy to or from the temperature sensor. The inner surfaces and the temperature sensor mount should be made from material of low thermal conductivity. Stainless steel is often used, as it has the lowest thermal conductivity of the commonly available metals (1 100 $W\ m^{-1}\ K^{-1}$).

When the temperature sensor is placed in concentric cylindrical shields, the orifice may comprise a significant amount of the sensor's total view. Moffat (1967) recommended that the length and diameter of the inner shield be selected so that the solid angle of view subtended by the open ends of the shield is less than 0.04 steradian (Sect. 2.1.2.4 for definition). This condition is satisfied in tubes if the sensor is recessed to a distance at least four times the tube diameter.

Instead of concentric shields, Sargent and Tanner (1967), Gay (1972), and Black and McNaughton (1971) have used a single shield covered with expanded foam insulation and an outer covering of aluminized mylar. This reduces the ventilation requirements and complexity of multiple shields.

Orientation of cylindrical shields is important. If horizontal, the intake should always be pointed north in the northern hemisphere to reduce reflection of solar radiation on the temperature sensor. The intake can be tilted slightly downward, but for the zone of influence to be compatible with other measurements such as wind, horizontal mounting appears best.

3.4.2   Air Temperature Gradients

Air temperature gradients are used in a number of atmospheric heat transfer analyses and they may also be of interest in ecological research. Generally, time averaged gradients for periods of 15 to 60 min are desired for Bowen ratio or aerodynamic analyses. Larger sensors with longer time constants are more easily constructed than the small, fast sensors; they also yield a smoothed output that is easier to average. Fast response sensors, on the other hand, require frequent sampling or integrating to develop time averaged readings. Aspirated sensors are common. They are generally more reliable under common wind and radiation conditions provided the aspiration rate is not too high.

Temperature gradients in the atmosphere may approach the dry adiabatic lapse rate of $0.01°C\ m^{-1}$. Thus, the accuracy required for gradient measurements may be more than an order of magnitude greater than that needed for ambient temperature. Radiation errors in gradients can be made small, provided the radiation errors in the two sensors are kept essentially the same. Consequently symmetry of sensors is desirable. One point that cannot be emphasized too strongly is that all temperature sensors used for air temperature profiles be identical in construction and exposure.

The question of desirable resolution may be important for the measurement of air temperature gradients. As a general guide, the measuring device should have enough sensitivity to resolve the variable measured into a hundred parts. The small values of the gradients thus require that the sensors selected have a high output and that sensitive measuring devices be available.

3.4.2.1   Vertical Sampling

Meteorological models require temperature gradients at two or more levels. The Bowen ratio energy balance analysis is based on time averaged differences, ΔT, measured at two heights above the surface and vapor pressure differences, Δe, measured over the same height interval. The aerodynamic analyses require time averaged temperature and vapor pressure differences between a number of levels.

Any of the temperature elements may be used for gradient measurements; thermopiles and diodes are used most frequently (Sargeant and Tanner, 1967; Fritschen, 1965; Black and McNaughton, 1971). When thermopiles are used for temperature differences, five to ten junctions are located at each level to increase the signal.

Two methods of measuring temperature differences are illustrated in Fig. 3.21 with 2-junction thermopiles. Any number of junctions could actually be used in the field. The two approaches are to measure temperature differences between successive levels [Fig. 3.21(a)] or to measure temperature differences with respect to one selected level [Fig. 3.21(b)].

The differences between successive levels can be summed to yield the total difference $(\Delta T_{1,4} = \Delta T_{1,2} + \Delta T_{2,3} + \Delta T_{3,4})$. To check the validity

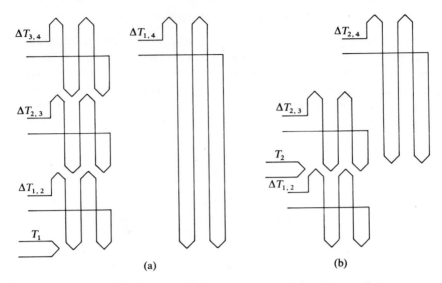

Figure 3.21(a) and **3.21(b)**. Two methods of using 2-junction thermopiles to measure temperature gradients; (a) differences between successive levels and (b) differences with respect to one level.

of the summed differences, a separate circuit can be run between the top and bottom levels to measure the total temperature difference directly [$\Delta T_{1,4}$ in Fig. 3.21(a)].

If the measurements are referenced to a common level [Fig. 3.21(b)], there is less concern about summing up errors with the temperature differences. This method requires additional thermocouple wire, and the multiple junctions at the common reference level may be bulky. The ice bath or standard temperature reference may be established at the reference level. If placed at the site of the measurements, the length of thermocouple lead wire between the sensors and the reference will be minimized.

In both configurations, a separate circuit should be run to measure the absolute temperature at one of the junctions. It is important to effectively insulate each of the electrical junctions.

These two methods illustrate the use of difference measurements in eliminating fixed errors or biases that are identical for each sensor. The biases may not be identical, however. This is more of a problem for wet-bulb temperatures (see Sect. 6.2.1) than for drybulb temperatures because of slight variability in water supply between the various sensors.

There are other ways to eliminate these variations. Periodically interchanging the temperature sensors will eliminate this variability. The temperature difference is averaged over enough cycles so that each sensor has spent equal time at both locations. This averages out any offsets that may exist between the sensors. Suomi (1957), Tanner (1963), Fritschen (1965), Sargeant and Tanner (1967), and Black and McNaughton (1971) have described designs of apparatus to be used for exchanging sensors. Rosenberg and Brown (1974) describe a variation; their system brings a number of psychrometers to a common level for periodic calibration.

Another way to minimize sensor differences is by traversing a single sensor, stopping at each level where a measurement is desired. However, gradients evaluated from this method require frequent samples and long period averages (Suomi, 1956). This method is not often used.

3.4.2.2   Spatial Horizontal Sampling

If vertical temperature and water vapor pressure profiles are measured well above the surface where the air is mixed and the velocities are higher, gradients are small and difficult to measure. Furthermore, the sensors may be at a level where vapor pressure, temperature, and wind are not in equilibrium with the surface unless the fetch is large enough to avoid divergence due to advection. In addition, effects of vertical stratification and buoyancy can be serious. On the other hand, measurements may be made near the surface where the gradients are steeper and more easily measured. Near the surface, a shorter fetch is required for vapor pressure, temperature, and wind profiles to come into equilibrium with the surface and thermal stratification is less important, but gradient sampling is affected by inhomogeneities of the surface. Spatial sampling becomes necessary to achieve representative values, Thus the experimenter must face either the problem of spatial sampling near the surface where the vertical gradients can be measured most easily, or instrument, fetch, and thermal stratification problems at higher levels.

Sensors have been traversed horizontally to obtain spatial sampling (Tanner, 1960; Fritschen and Van Bavel, 1963). Even when spatial sampling is successful, however, there is no assurance that either the energy balance or the aerodynamic methods will be reliable over inhomogenous surfaces. For example, widely spaced rows, skip rows, very patchy crops, or wet and dry surfaces represent patchy surfaces where the heat and water vapor pressure sinks and sources are separated from the momentum sinks. Further studies are needed to determine that micrometeorological measurements under such conditions can be applied to analyses of heat and mass transfer.

The aspiration of the sensors may affect the measurements as removal of air changes the temperature structure. This effect will be the largest under calm light winds; as the wind velocity increases, the sample is drawn from a progressively smaller region about the intake orifice. Also, the gradient in natural wind may affect the symmetry of the area sampled. For this reason, carefully designed unaspirated thermometers may be superior to the more common aspirated designs. The benefits of natural ventilation during periods of calm light winds are offset by the radiation errors in the exposed sensors or unaspirated shields. In any event, large aspiration rates could give an error at most wind speeds.

The efficiency of horizontally versus vertically oriented aspirated shields has also been questioned. A general preference exists for the horizontal orientation, but the actual orientation of intakes depends on the intended use of the data. When the temperature differences are to be used with vapor pressure differences, as in the Bowen ratio method, for example, there seems to be little preference of one orientation over the other. It is important that the same air is analyzed for temperature and vapor pressure. When temperature profiles are used with wind profiles, as in aerodynamic flux analyses, the temperature and velocity measurements must represent the same height. Knowing the sampling height is most important at the lower levels where the gradients are large and sampling is done at smaller height intervals. If errors exist, either due to orientation of aspirated shield intakes or due to aspiration, it would be more serious with aerodynamic measurements than with the Bowen ratio method.

3.4.3   Temperature in Plant Canopies or Near the Surface

Measurement of air temperature either within the plant air canopy layer or near the surface is complicated by several problems. First, if a nonaspirated sensor with a large shade is used, the shadow of the sensor will affect the radiation exchange at the underlying surfaces; this will affect the temperature. On the other hand, if air is aspirated through small shields, the air intake may create velocities larger than those found in the canopy or near the soil surface, and this may cause an error. It is apparent that the small unaspirated shields described by Long (1957) and the unshielded oscillating sensor described by DeWit (1954) offer advantages. Circumstances of measurement, natural wind velocities, amount of radiation, and space available are factors that will affect the choice of sensor. If the measurement is not too near the plant or soil surface, miniature shields with low aspiration rates can be used, provided the sensors are traversed so that the effect of shadow on air intake is spread over a large area. Traversing the sensor offers the additional advantage of spatial sampling in the near-surface area where spatial inhomogeneities exist.

3.5   Soil Temperature Measurements

Soil temperature can be measured by any of the temperature sensors. The problems associated with soil temperature measurements are twofold: (1) placement of the sensors without disturbing the soil and (2) horizontal variability of surface cover and soil properties. Portman (1957) described a convenient method for inserting temperature sensors into an undisturbed

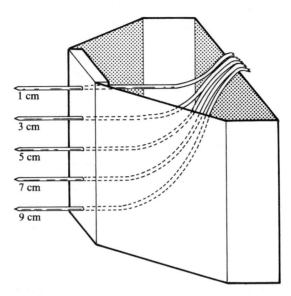

Figure 3.22. Sketch of soil thermocouple installation at 1, 3, 5, 7, and 9 cm depths.

fine-textured soil. A modification used by the author is shown in Fig. 3.22. Brass tubes (2 mm diameter × 15 cm) are inserted through the reference board (2 cm thick × 4 cm wide) at the desired depths. The ends of the brass tubes away from the board are pointed and sealed. Temperature sensors (i.e., 30-gauge B & S thermocouples) are sealed into the brass tubes, and the assembly is placed in the soil by pushing the sharpened brass tubes into a smoothed side of a triangular pit. The top of the board is even with the soil surface and serves as a depth reference. The pit is then back filled.

Changes in the average temperature of a layer, rather than individual temperatures, are needed for soil heat storage calculations. Thermocouples at various depths can be connected in parallel to give the average temperature of the layer that they define. Several installations can be connected in parallel to obtain the average temperature at several locations (Fritschen, 1965). Suomi (1957a) used resistance thermometers, 5 cm long, installed vertically in the soil to obtain average layer temperatures. In addition, he connected 12 such resistance elements in series to obtain a spatial average. Tang et al. (1974) placed five diodes in series in the soil to obtain the average temperature of a layer.

A technique for inserting a temperature sensor into an undistributed soil at a greater depth consists of cementing an insulated thermocouple into a cavity made in a small piece of sharpened rod. A pointed rod of the desired length is also inserted into the cavity and is used to place the thermocouple at the desired depth in a hole previously made with a small steel rod. The rod is removed leaving the thermocouple in place. The thermocouple may be removed later if desired by pulling the lead.

Bibliography

American Society for Testing and Materials (1974). Manual on the Use of Thermocouples in Temperature Measurement. Am. Soc. Test. Meter. Spec. Tech. Publ. **470A**.

Benedict, R. P. (1977). Fundamentals of Temperature, Pressure, and Flow measurements, 2nd ed. Wiley, New York. 517 pp.

Finch, D. I. (1962). General Principles of Thermoelectric Thermometry. Leeds & Northrup Tech. Pub. 01.1000.

Middleton, W. E. K., and A. F. Spilhaus (1960). Meteorological Instruments. Univ. Toronto, Toronto, Canada. 286 pp.

Tanner, C. B. (1963). Basic Instrumentation and Measurements for Plant Environment and Micrometeorology. Dept. Soils Bull. 6. Univ. Wisconsin, Madison, Wisc.

Literature Cited

Barton, L. E. (1962). Measuring temperature with diodes and transistors. *Electronics* **35**:38–40.

Benedict, R. P. (1977). Fundamentals of Temperature, Pressure, and Flow measurements, 2nd ed. Wiley, New York. 517 pp.

Black, T. A., and K. G. McNaughton (1971). Psychrometric apparatus for Bowen-ratio determination over forests. *Boundary-Layer Meteorol.* 2:246–254.

Callendar, H. L. (1887). On the construction of platinum temperature sensors. *Philos. Trans. R. Soc. London.* **178**:161–230.

Dimick, R. C., and G. J. Trezek (1963). Photodiodes as a sensitive temperature probe. *Rev. Sci. Instrum.* **34**:981–983.

DeWit, C. T. (1954). An oscillating psychrometer for micrometeorological purposes. *Appl. Sci. Res.* **4**:120–126.

Fritschen, L. J. (1965). Accuracy of evapotranspiration determinations by the Bowen-ratio methods. *Bull. Intl. Assoc. Sci. Hydrol.* **2**:38–48.

Fritschen, L. J., and C. H. M. van Bavel (1963). Experimental evaluation of models of latent and sensible heat transport over irrigated surfaces. Committee for evaporation. Intl. Assoc. Sci. Hydrol. Pub. No. 62, p. 159–171.

Fuchs, M., and C. B. Tanner (1965). Radiation shields for air temperature thermometers. *J. Appl. Meteorol.* **4**:544–547.

Gay, L. W. (1972). On the construction and use of ceramic wick thermocouple psychrometers. In: Brown, R. W. and B. P. VanHaveren (eds.). Psychrometry in Water Relations Research. p. 251–258. Utah Agricultural Experiment Station, Logan. 318 pp.

Hinshaw, R., and L. J. Fritschen (1970). Diodes for temperature measurement. *J. Appl. Meteorol.* **9**:530–532.

Hickson, V. M. (1940). The welding of thermocouple junctions. *J. Sci. Instrum.* **17**:182–186.

Kaye, G. W. C. and T. H. Laby (1973). Tables of Physical and Chemical Constants. Longmans, London. 14th ed. 386 pp.

King, W. J. (1943). Measurement of high temperature in high velocity gas streams. *Am. Soc. Mech. Eng. Pap.* **65**:421–431.

Le May, J. (1958). More accurate thermocouples with percussion welding. *J. Instrum. Soc. Am.* **5**:42–45.

Long, D. F. (1957). Instruments for micrometeorology. *Q. J. Roy. Meteorol. Soc.* **83**:202–214.

Lopushinsky, W. (1971). An improved welding jig for Peltier thermocouple psychrometers. *Soil. Sci. Soc. Am. Proc.* **35**:149–150.

McNamara, A. G. (1962). Semiconductor diodes and transistors as electrical thermometers. *Rev. Sci. Instrum.* **33**:330–333.

Moffatt, E. M. (1967). Gas Temperature. In: Measurement Engineering. Stein Engineering Service, Phoenix, Arizona, 745 pp.

National Physical Laboratory (1969). The International Practical Temperature Scale of 1968. Ministry of Technology, HMSO, London.

Powell, R. L., W. J. Hall, C. H. Hyink Jr., L. L. Sparks, G. W. Burns, M. G. Scroger, and H. H. Plumb (1974). Thermocouple Reference Tables Based on the IPTS-68. *Natl. Bur. Stand. U.S. Monogr.* **125**.

Portman, D. J. (1957). Soil thermocouples. p. 19–20. In: Exploring the Atmosphere's First Mile. Pergamon, New York. 376 pp.

Riddle, J. L., G. T. Furukawa, and H. H. Plumb (1973). Platinum Resistance Thermometry. *Nat. Bur. Stand. U.S. Monogr.* **126**.

Roeser, W. F., and S. T. Louberger (1958). Methods of Testing Thermocouples and Thermocouple Materials. *Nat. Bur. Stand. U.S. Circ.* **50**.

Rosenberg, N. J., and K. W. Brown (1974). "Self-checking" pyschrometer system for gradient and profile determinations near the ground. *Agric. Meteorol.* **2**:215–226.

Sargeant, D. H. (1965). Note on the use of junction diodes as temperature sensors. *J. Appl. Meteorol.* **4**:644–646.

Sargeant, D. H., and C. B. Tanner (1967). A simple psychrometric apparatus for Bowen-ratio determinations. *J. Appl. Meteorol.* **6**:414–418.

Stover, C. M. (1960). Method of butt welding small thermocouple 0.001 to 0.010 inch in diameter. *Rev. Sci. Instrum.* **31**:605–608.

Suomi, V. E. (1956). Energy Budget Studies at the Earth's Surface and Development of the Sonic Anemometer for Power Spectrum Analysis. Univ. of Wisc., Dept. of Meteorol. (Rpt. AFCRC TR-56-273).

Suomi, V. E. (1957a). Soil temperature integrators, p. 24. In: Exploring the Atmosphere's First Mile. Pergamon, New York. 376 pp.

Suomi, V. E. (1957b). Double-psychrometer lift apparatus. p. 183–187. In: Exploring the Atmosphere's First Mile. Pergamon, New York. 376 pp.

Tang, P. A., K. G. McNaughton, and T. A. Black (1974). Inexpensive diode thermometry using integrated circuit components. *Can. J. For. Res.* **4**:250–254.

Tanner, C. B. (1960). Energy balance approach to evapotranspiration from crops. *Soil Sci. Soc. Am. Proc.* **24**:1–9.

Tanner, C. B. (1963). Basic Instrumentation and Measurements for Plant Environment and Micrometeorology. Dept. Soils Bul. 6. Univ. of Wisconsin, Madison, Wisc.

Trolander, H. W. (1972). The current state of electrical thermometry for biological applications. p. 2035–2050. In: Temperature, Its Measurement and Control in Science and Industry. H. H. Plumb (ed.). Instrument Soc. of Am., Pittsburgh, Pa. 2383 pp.

Chapter 4

Soil Heat Flux

4.1   Soil Heat Flux Transducer

Soil heat flux is sometimes ignored because its magnitude is often small relative to the other terms in the energy balance equation. The daily and yearly sums are normally near zero, but daily values can be significant when the soil is heating in the spring or cooling in the fall. Hourly values of the soil heat flux can be large under dry bare soil conditions. Omission of soil heat flux measurements from hourly energy budget analyses will introduce a phase error in the evaluation even though the 24-hour totals will remain unchanged.

Soil heat flux can be calculated from temperature gradients if soil thermal conductivity is known, or from changes in temperature if the heat capacity is known. Values of thermal conductivity and heat capacity are not often available. The values also vary with soil moisture content, so a direct measurement of the soil heat flux is often the simplest approach to follow.

Direct measurements can be made with relatively simple transducers. The heat flow through the soil heat flux transducer (G_T) is related to the temperature difference across the transducer,

$$G_T = \frac{\lambda(T_t - T_b)}{l} \tag{4.1}$$

where λ is the thermal conductivity, l is the thickness of the transducer, and T_t and T_b are the temperatures at the top and the bottom of the transducer. Thermopiles usually used in the transducer develop an emf in response to ΔT that is proportional to the heat flux across the plate (G_T). The flux across the plate is equal to that in the surrounding soil when its conductivity equals that of the soil. The bobbin on which the thermopile is wound should therefore, have a thermal conductivity similar to the medium in which it is placed. The thermal conductivity of soil depends on moisture and is constantly changing. As a necessary compromise, the plate material should have a thermal conductivity equal to that of the soil, under average conditions. It is also important that the bobbin be dimensionally stable and not absorb water. Glass, bakelite, and plastic resins have been used for bobbin material. The complete transducer assembly should be encapsulated rather than laminated to prevent deterioration in the warm moist soil environment.

A thermopile will generate a thermoelectric potential, D, that is related to the temperature difference by

$$D = NC(T_t - T_b), \tag{4.2}$$

where N is the thermoelectric power of the thermopile material and C is a proportionality constant. Combining Eqs. (4.1) and (4.2) yields

$$G_T = \left(\frac{\lambda}{l}\right)\left(\frac{D}{NC}\right) = gD, \tag{4.3}$$

where g is the calibration constant relating G_T to D.

The theory of proper heat flux transducer design has been discussed by Philip (1961). For thin circular transducers, the ratio of heat flux through the transducers to the flux density through the medium, G_G, is given by

$$\frac{G_T}{G_G} = \frac{1}{[1 - 1.92(l/d)(1 - 1/\varepsilon)]}, \tag{4.4}$$

where l is the thickness, d is the diameter, and ε is the ratio of transducer conductivity to medium conductivity.

To minimize errors, heat flux transducers should be (1) made as thin as possible, (2) calibrated in a medium with thermal conductivity similar to the soil of intended use, and (3) made of material with appropriate thermal conductivity. For example, sand and clay soils with water contents of 8–20% by volume have thermal conductivities that range from 0.8 to 2.2 W(m K)$^{-1}$. The heat flux transducers constructed from plastics, glass, and bakelite. have thermal conductivities of about 0.17 W(m K)$^{-1}$ and could have a maximum error as great as 44%. However, poor contact with the soil could result in larger errors. If air gaps were present at either the top or bottom, the error could be as great as 54% (Philip, 1961). Hatfield (1945) described a transducer consisting of a single thermocouple constructed from copper tellurium. Its larger thermal conductivity would reduce the error discussed above. However, the sensitivity is low, and the transducer needs electrical insulation, which reduces its thermal conductivity.

4.2   Soil Heat Flux Measurements

Soil heat flux transducers should not be located close to the soil surface because the transducers could impede root development and will resist the natural flow of moisture that may result in an unnaturally dry layer above to the transducer. In addition, soil above the transducer may crack exposing it to direct insolation. As a practical compromise, Tanner (1960) suggested that the transducer be located 5–10 cm below the soil surface. The change in energy storage of the layer of soil above the transducer is estimated from the heat capacity equation and supplementary measurements in temperature within the layer.

When the soil heat flux transducer is located at a depth Δz, the energy flow at the surface is given by

$$G_0 = -\lambda\left(\frac{\Delta T}{\Delta z}\right) - C\,\Delta z\left(\frac{\Delta \overline{T}}{\Delta t}\right). \tag{4.5}$$

The first term is soil heat flux and is measured with the transducer. The second term is the change in energy storage above the transducer where $\overline{T}$ is the average temperature of the layer and C is the heat capacity of the layer.

The heat capacity of the soil can be estimated from

$$C = X_m C_m + X_o C_o + X_w C_w + X_a C_a, \tag{4.6}$$

where X is the volume fraction and C the heat capacity of minerals (m), organic matter (o), water (w), and air (a). Using nominal values for the heat capacities of the constituents, a soil having 50% minerals, 10% organic matter, and 40% water has a heat capacity of

$$C = (0.5)(2.0) + (0.1)(2.5) + (0.4)(4.2) = 2.9 \text{ MJ/(m}^3 \text{ K})^{-1}. \tag{4.7}$$

The air term was ignored because its heat capacity is very small relative to the other terms. The same soil with 10% water (a very dry condition) would have a heat capacity of 1.7 MJ(m^3 K)$^{-1}$. Heat capacity is a function of the water holding capacity, so the range in heat capacity is greater for finer materials such as peat and clay and less for coarser materials such as sands and gravel. As is evident from Eq. (4.5), a given relative error in estimation of C will result in a smaller error in G_0.

4.3   Sampling Requirements

The number of sampling locations required varies with the experimental conditions. Fewer locations are required under bare soil, uniform dense grass, or other uniform dense vegetation than under heterogeneous conditions. If the prime purpose of a study is the evaluation of the energy balance, more effort is expended evaluating the other terms, and sampling soil heat flux is dependent on time and money available. No general rules are available on the number of samples required. However, at least three sample locations would be a reasonable number under good conditions.

4.4   Calibration of Heat Flux Transducers

The various methods used to calibrate heat flux transducers can be grouped into either conduction or radiometric methods.

4.4.1   Conduction Methods

The conduction methods can be subdivided depending on whether divergence or convergence is present. These methods will be referred to as 1 and 2. These conditions occur whenever the thermal conductivity of the medium differs from that of the transducer. Hatfield and Wilkins (1950) and Fuchs and Tanner (1968) calibrated the heat flux transducer in a medium with smaller thermal conductivity than that of the transducer. The heat flux through the medium is estimated from the power dissipated by a heater which can be equipped with a guard ring. The thermal conductivity of the medium is not needed for the calibration.

Fritschen (unpublished) has located the heat flux transducer in a slurry of glass beads and water whose thermal conductivity is slightly larger than that of the transducer. The glass bead mixture was placed between a hot and a cold bath to establish a heat flux (method 3). The heat flux through the mixture is calculated from the thermal conductivity of the mixture, its thickness, and the temperature gradient.

Schwerdtfeger (1970) and Idso (1972) describe methods (number 4 and 5) that are based on conductivity without divergence. Schwerdtfeger (1970) mounted the heat flux meter directly between two tanks of water and measured the temperature gradient across the transducer. This technique requires knowledge of the thermal conductivity of the transducer under test. The method of Idso is similar to that of Fuchs and Tanner (1968) except that the transducer is located in a medium of similar thermal conductivity. The heat flux through the calibration medium is estimated from the power dissipated by the heater.

Deacon (1950) described a calibration device similar to that of Hatfield and Wilkins (1950) except that the transducer is placed directly on top of a copper cylinder containing a heater (method 6). A thin layer of vaseline insures good thermal contact between the transducer and the heated cylinder. Heat flux is determined by the power dissipation of the heater. Since one side of the transducer is exposed to environmental conditions, the heat flux through the transducer is subject to variations of the environmental conditions, mainly wind.

The authors have placed a transducer between two aluminum rods of the same diameter as the transducer (method 7). An electrical heater is located in the end of one of the rods. Both rods are insulated so that the energy from the heater travels from the first rod through the transducer and is dissipated by convection from the end of the other rod. The temperature gradient across the transducer is measured with thermocouples located in the aluminum rods.

4.4.2   Radiometric Methods

Two radiometric methods of heat flux transducer calibration have been described recently. Both methods establish heat flow through the transducer by placing it in a thermal radiation field. The intensity of the radiation fluxes and the emissivities of the transducer surfaces must be known. Schwerdtfeger (1970) used a steady state method (method 8) in which convection errors were eliminated by evacuating the chamber. Idso (1972)

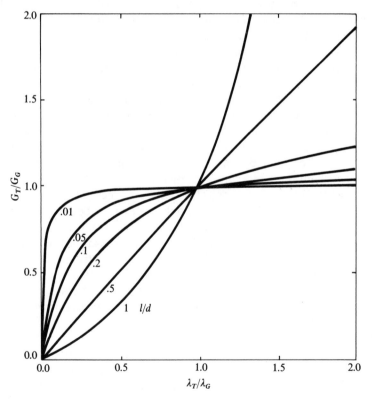

Figure 4.1. Relation between the ratio of flux density through the transducer G_T to that through the medium G_G as a function of the ratio of the thermal conductivity of the transducer λ_T to the thermal conductivity of the medium λ_G for transducers having various thickness, l, to diameter, d, ratios.

described a non-steady-state method (method 9) of calibrating heat flux transducers. He assumed that convection errors were negligible, but took no specific steps to guard against them.

4.4.3   Comments on Calibration Methods

Equation (4.4) is plotted in Fig. 4.1 for transducers of various thickness to diameter ratios, l/d. For example, consider a transducer of $l/d = 0.05$ and a thermal conductivity of 0.17 W(m K)$^{-1}$. Assume that this transducer is to be used in sand with 20% water by volume and thermal conductivity of 2.2 W(m K)$^{-1}$ ($\lambda_T/\lambda_G = 0.08$, $G_T/G_G = 0.47$). If the transducer were calibrated in a medium of 2.0 W(m K)$^{-1}$ (method 3), the calibration factor would be correct for that medium. As the soil dried to 8% water by volume (1.8 W(m K)$^{-1}$, $\lambda_T/\lambda_G = 0.09$, $G_T/G_G = 0.52$) the calibration factor would be too large by about 1%.

Suppose that the transducer was calibrated by methods 5, 6, 7, 8, and 9, or by method 2 in a medium of 0.17 W(m K)$^{-1}$ conductivity. If the transducer was then placed in the above soil conditions without correcting for the conductivity ratio, the error in measured heat flow would be 53% with 20% water and 48% with 8% water.

It is obvious in Fig. 4.1 that errors would be the least if thin transducers of high thermal conductivity were used. The signal produced by this thermopile design would be extremely small and difficult to measure because l and ΔT would be small. The most satisfactory approach appears to be that of correcting the measured heat flux, based on our knowledge of the calibration technique used and the ratio of conductivities of the transducer and the medium.

Bibliography

Lettau, H. L. and B. Davidson (1957). Exploring the Atmosphere's First Mile, Vol. 1. Pergamon, New York. 376 pp.

Sellers, W. D. (1965). Physical Climatology. Univ. of Chicago, Chicago, Illinois. 272 pp.

Wijk, W. R. van (1963). Physics of Plant Environment. North–Holland Publishing Co., Amsterdam. 382 pp.

Literature Cited

Deacon, E. L. (1950). The measurement and recording of the heat flux into the soil. Q. J. Roy. Meteorol. Soc. 76:479–483.

Fuchs, M. and C. B. Tanner (1968). Calibration and field test of soil heat flux plates. Soil Sci. Soc. Am. Proc. 32:326–328.

Hatfield, H. S. (1945). An improved means of measuring flow of heat. Patent Specifications. The Patent Office, 25 Southampton Building, London, W.C. 2, England.

Hatfield, H. S. and F. J. Wilkins (1950). A new heat-flow meter. *J. Sci. Instrum.* **27**:1–3.

Idso, S. B. (1972). Calibration of soil heat flux plates by a radiation technique. *Agric. Meteorol.* **10**:467–471.

Philip, J. R. (1961). The theory of heat flux meters. *J. Geophys. Res.* **66**:571–579.

Schwerdtfeger, P. (1970). The measurement of heat flow in the ground and the theory of heat flux meters. Cold Regions Research and Engineering Laboratory DA Task 4A062112A89401. Corps of Engineers, U.S. Army, Hanover, New Hampshire.

Tanner, C. B. (1960). Energy balance approach to evapotranspiration from crops. *Soil Sci. Soc. Am. Proc.* **24**:1–9.

Chapter 5

Radiation

Radiation is the process by which energy can be transferred from one body to another through electromagnetic waves in the absence of an intervening medium. If an intervening medium is present, it must be at least partially transparent in order for the radiant energy transfer to take place. Energy is continuously radiated from all substances that are above absolute zero (0 K) in temperature. If a black body is warmer than its environment, it radiates more energy than it receives and the environment receives more energy than it radiates. There is, thus, a net transfer of energy from the warm black body to the cooler environment. The basic problem is to determine the direction and rate of this net energy transfer.

Emission of electromagnetic energy results in a decrease in the energy level of the molecule, and absorption results in an increase in the energy level. However, there are some radiation phenomena that do not maintain equilibrium between the thermal motion of the molecules and the net gain or loss of radiant energy. Examples are the emission of energy by radioactive decay and the fluorescence of a phosphor under bombardment by electrons. Radiation of this nature is not important for measurement techniques considered in this work, so the term radiation will refer only to that which is thermal in origin.

5.1 Radiation in Various Wave Bands

Certain wavelength bands on the scale of electromagnetic radiation have become associated with common names, usually because of a uniform characteristic that applies to the entire band. The names and outer limits of these bands are summarized in Table 5.1. For convenience, the radiation

Table 5.1. The electromagnetic spectrum.

Type of radiation	Range of wavelength (μm)	Range of frequency (s^{-1})
Cosmic rays, gamma rays, x-rays, etc.	≤ 0.001	$>0.3 \times 10^{18}$
Ultraviolet	0.001 to 0.4	10^{15} to 0.3×10^{18}
Visible	0.4 to 0.8	0.4×10^{15} to 10^{15}
Near infrared	0.8 to 4	80×10^{12} to 0.4×10^{15}
Far infrared	4 to 100	3×10^{12} to 80×10^{12}
Microwave	100 to 10×10^6	30×10^6 to 3×10^{12}
Radio	$\geq 10 \times 10^6$	$\leq 30 \times 10^6$

from the sun ($\lambda < 4 \mu$m) and natural bodies ($\lambda > 4 \mu$m) are further classified as shortwave and longwave radiation.

The basic laws governing the transfer of radiation were reviewed briefly in Sect. 2.1.2.4. The Stefan–Boltzmann law [Eq. (2.18)] describes the emission of thermal radiation by a body as $E = \varepsilon \sigma T^4$ (W m^{-2}) where σ, the Stefan–Boltzmann constant, is equal to 56.697×10^{-9} W m^{-2} K^{-4}. The law shows that the broad band emission over all wavelengths is proportional to the fourth power of the absolute temperature (K) of the body. The total emitted radiant energy for black bodies ($\varepsilon = 1.0$) at various temperatures is tabulated in Table 5.2.

The emitted energy is tabulated in Table 5.2 in SI units (W m^{-2} and also in the common but no longer approved units of cal cm^{-2} min^{-1}. A similar unit is the langley (ly), which honored Samuel Pierpont Langley who contributed significantly to studies of the solar constant during his tenure as Secretary of the Smithsonian Institution at the turn of the century. The unit was proposed by Alrich et al. (1947) as 1 ly = 1 cal cm^{-2}. As a matter of historical interest, the definition was based on the "15°C calorie," now set equal to exactly 4.186 8 J. However, common usage has placed many radiation measurements on a scale defined by the thermochemical calorie, which is equal to exactly 4.1840 J in the SI conversion system. The difference between the definition and the common use is less than 0.1%, but the best correspondence appears to be 1 ly = 1 (thermochemical) cal cm^{-2} = 41.84 kJ m^{-2}. Future use of the langley (as well as the calorie) will be limited, however, due to acceptance of SI.

The energy content of radiation, particularly that of photosynthetically active radiation in the visible range, is sometimes discussed in terms of the discrete photons, or quanta, of electromagnetic energy at a specific wavelength. The energy, Q, in a photon or quantum of radiation is related to the frequency, v, of the wave by Planck's constant, h,

$$Q = hv = \frac{hc}{\lambda} = hcn, \qquad (5.1)$$

Table 5.2. Blackbody radiation (σT^4) in W m^{-2} (upper line) and cal cm^{-2} min^{-1} (lower line).

°C	0	1	2	3	4	5	6	7	8	9
−50	140	143	145	148	151	153	156	159	162	164
	0.201	0.205	0.208	0.212	0.216	0.220	0.224	0.228	0.232	0.236
−40	167	170	173	176	179	182	185	188	191	195
	0.240	0.244	0.248	0.252	0.257	0.261	0.265	0.270	0.274	0.279
−30	198	201	204	208	211	215	218	222	225	229
	0.284	0.288	0.293	0.298	0.303	0.308	0.313	0.318	0.323	0.328
−20	232	236	240	244	247	251	255	259	263	267
	0.333	0.338	0.344	0.349	0.355	0.360	0.366	0.372	0.377	0.383
−10	271	276	280	284	288	293	297	301	306	311
	0.389	0.395	0.401	0.407	0.413	0.420	0.426	0.432	0.439	0.445
0	315	320	324	329	334	339	344	349	354	359
	0.452	0.458	0.465	0.472	0.479	0.486	0.493	0.500	0.507	0.514
10	364	369	374	380	385	390	396	401	407	412
	0.522	0.529	0.537	0.544	0.552	0.559	0.567	0.575	0.583	0.591
20	418	424	430	435	441	447	453	460	466	472
	0.599	0.608	0.616	0.624	0.633	0.641	0.650	0.659	0.668	0.676
30	478	485	491	497	504	511	517	524	531	538
	0.685	0.695	0.704	0.713	0.722	0.732	0.741	0.751	0.761	0.771
40	544	551	559	566	573	580	587	595	602	610
	0.781	0.791	0.801	0.811	0.821	0.832	0.842	0.853	0.863	0.874
50	617	625	633	641	649	657	665	673	681	689
	0.885	0.896	0.907	0.918	0.930	0.941	0.953	0.964	0.976	0.988
60	698	706	714	723	732	740	749	758	767	776
	1.000	1.012	1.024	1.036	1.049	1.061	1.074	1.087	1.100	1.113
70	785	794	804	813	822	832	842	851	861	871
	1.126	1.139	1.152	1.165	1.179	1.193	1.206	1.220	1.234	1.248
80	881	891	901	911	921	932	942	953	963	974
	1.263	1.277	1.292	1.306	1.321	1.336	1.351	1.366	1.381	1.396
90	985	996	1007	1018	1029	1040	1052	1063	1075	1086
	1.412	1.428	1.443	1.459	1.475	1.491	1.508	1.524	1.541	1.557
100	1098	1110	1122	1134	1146	1158	1170	1183	1195	1208
	1.574	1.591	1.608	1.625	1.643	1.660	1.678	1.696	1.714	1.732
110	1221	1233	1246	1259	1272	1286	1299	1312	1326	1340
	1.750	1.768	1.787	1.805	1.824	1.843	1.862	1.881	1.901	1.920
120	1353	1367	1381	1395	1409	1423	1438	1452	1467	1482
	1.940	1.960	1.980	2.000	2.020	2.040	2.061	2.082	2.103	2.124
130	1496	1511	1526	1541	1557	1572	1587	1603	1619	1635
	2.145	2.166	2.188	2.210	2.231	2.253	2.276	2.298	2.320	2.343
140	1650	1667	1683	1699	1715	1732	1749	1765	1782	1799
	2.366	2.389	2.412	2.435	2.459	2.483	2.506	2.530	2.555	2.579
150	1816	1833	1851	1868	1886	1904	1922	1939	1958	1976
	2.604	2.628	2.653	2.678	2.703	2.729	2.754	2.780	2.806	2.832

where c is the speed of light (299.8×10^6 m s^{-1}), λ is wavelength (m), n is wavenumber (m^{-1}) and h is equal $0.662\,62 \times 10^{-33}$ J s. The energy is green light, for example, with average wavelength $\lambda = 550$ nm would be [from Eq. (5.1)] $(0.662\,62 \times 10^{-33}$ J s$)(299.8 \times 10^6$ m s$^{-1})/550$ nm $= 0.36\,1 \times 10^{-18}$ J. Since this number is small, it is multiplied by the number of photons in a mole (602×10^{21}), which is called an Einstein, to obtain the energy of photons in green light as (602×10^{21} Einstein^{-1}) $(0.361 \times 10^{-18}$ J$) = 217.435$ kJ Einstein^{-1}. If the radiant flux density from the sun, for example, were 500 W m^{-2}, the photon flux of green light would be $(500$ J s^{-1} m$^{-2})/(217.435$ kJ Einstein$^{-1}) = 2.3 \times 10^{-3}$ Einstein m^{-2} s^{-1}.

5.2   Methods of Radiation Measurement

Radiant energy may be detected in three ways: (1) by the rise of temperature of the small receiving surface, (2) by the response of a photoelectric cell, and (3) by photochemical methods. The thermal sensitive devices may generate an electrical signal; however, they should not be considered as photoelectric devices. The thermal sensitive devices are the most commonly used to measure radiation.

Thermal sensitive devices may take the form of temperature sensors such as a resistance element, a thermopile, or a distillometer. It is important to understand the principle of operation of thermal sensitive devices, whether they utilize resistance elements or thermopiles. Therefore, a brief review of the principle of operation of a net radiometer is presented because this information is applicable to other types of sensors.

In general, radiometers produce a signal because of a differential absorption of radiation. The principle of operation of radiometers is essentially the same whether the differential absorption of radiation is between a black and a white surface, two black surfaces, or a blackened surface and the base or housing of the instrument. For example, let us consider a blackened plate containing a thermal transducer placed in a radiation field. The energy budget of the top surface t of the blackened plate is given by

$$\alpha Q_t = \varepsilon \sigma T_t^4 + h(T_t - T_a) + \frac{\lambda}{l}(T_t - T_b), \qquad (5.2)$$

where α is the absorption coefficient, Q_t is the radiation flux density, ε is the emissivity, σ is the Stefan–Boltzmann constant, h is the thermal convection coefficient, λ is the thermal conductivity, and l is thickness of the transducer, T_a is air temperature, T_b is the temperature of the bottom of the plate. Similarly, the energy balance for the bottom surface, b, is

$$\alpha Q_b = \varepsilon \sigma T_b^4 + h(T_b - T_a) - \frac{\lambda}{l}(T_t - T_b). \qquad (5.3)$$

Net radiation, Q^*, is obtained by subtracting Eq. (5.3) from Eq. (5.2),

$$Q^* = \alpha(Q_t - Q_b) = \varepsilon\sigma(T_t^4 - T_b^4) + h(T_t - T_b) + \frac{2\lambda}{l}(T_t - T_b). \quad (5.4)$$

If the terms involving the thermal convection were eliminated (to be discussed later), the above equation would become

$$Q^* = \varepsilon\sigma(T_t^4 - T_b^4) + \frac{2\lambda}{l}(T_t - T_b). \quad (5.5)$$

The heat flow through the sensing element may be expressed as

$$G = \frac{\lambda}{l}(T_t - T_b). \quad (5.6)$$

It may be detected by the aid of a thermopile. The temperature difference between the hot and cold junctions of the thermopile produce a signal depending on the number, n, of junctions involved in the temperature gradient. The signal is related to the overall temperature difference $(T_t - T_b)$ by the thermoelectric power, N. A thermoelectric potential, D, is set up that is related to the temperature difference by $D = NC(T_t - T_b)$ where C is a proportionality constant.

By using the approximation that $T_t^4 - T_b^4 \simeq 4T_t^3(T_t - T_b)$ in the Eq. 5.5 we have

$$Q^* = \left(4\varepsilon\sigma T_t^3 + \frac{2\lambda}{l}\right)(T_t - T_b), \quad (5.7)$$

or

$$Q^* = \left(4\varepsilon\sigma T_t^3 + \frac{2\lambda}{l}\right)\left(\frac{D}{NC}\right) = gD, \quad (5.8)$$

where g is the calibration constant that depends on the temperature, emissivity, conductivity, and thickness of the thermal transducer. The magnitude of $4\varepsilon\sigma T_t^3$ is small compared to $2\lambda/l$ and can be disregarded.

The thermal convection term can be reduced or eliminated by enclosing the sensor portion of the thermal transducer of shortwave radiation measuring radiometers with a windshield or windshields of quartz and glass and in other radiometers with windshields of plastic or by directing a ventilation stream on both surfaces. The quartz and glass radiation windshields are generally transparent to wavelengths below approximately 4 and 3 μm, respectively.

Instruments commonly used by researchers for measuring radiation can be classified into three main classes: (a) those sensitive to shortwave radiation, (b) those sensitive to longwave radiation, and (c) those sensitive to both longwave and shortwave radiation. The needs of the user and the research problem will generally dictate the type of instrument required.

Shortwave radiation measurements most commonly used are *global solar radiation, K↓*, received on a horizontal surface, and *reflected solar radiation, K↑*. Global solar radiation includes both radiation received direct from the solid angle of the sun's disc and radiation that has been scattered or diffusely reflected in transversing the atmosphere. Occasionally global solar radiation is separated into its two components, the *direct solar beam, S* and the sky or *diffuse radiation, D*. Upward shortwave radiation consists of solar radiation reflected by the earth's surface and diffused by the atmospheric layer between the ground and the point of observation. The sum of the global solar radiation minus the reflected solar radiation is called *net solar radiation, K**.

Total radiation, Q, consists of shortwave and longwave radiation. When one is considering the downward flux, total radiation, $Q\downarrow$, would include the global solar radiation, $K\downarrow$, and the *longwave atmospheric radiation* mainly emitted by the atmosphere, $L\downarrow$. If upward radiation is considered, total radiation, $Q\uparrow$, would include the reflected solar radiation, $K\uparrow$, the terrestrial radiation, and the longwave atmospheric radiation between the surface and the point of measurement, $L\uparrow$. The sum of the total radiation fluxes down minus the total radiation fluxes up is referred to as *net radiation, Q**.

Instruments used for measuring the different radiation fluxes have special names. An instrument used for measuring radiation is called a radiometer. A *pyranometer* is used to measure global solar radiation. The *pyrheliometer* is used exclusively for measuring the direct beam solar radiation. Unfortunately, the word pyrheliometer has been associated with global solar radiation. A *pyrradiometer* is used for measuring total radiation, both shortwave and longwave, and a *pyrgeometer* is used for measuring the longwave atmospheric radiation. A *net pyrradiometer* is used for measuring net radiation. A more common name for this instrument, which was formerly called a balance meter, is net radiometer.

5.3   Radiation Instruments

Excellent descriptions of the most commonly used radiation instruments have been presented by I. G. Y. (1958), World Meterological Organization (1971), Reifsnyder and Lull (1965), Robinson (1966), and Coulson (1975), among others. These sources plus more recent information have been utilized for the following discussion.

The WMO Commission for Instruments and Methods of Observations has classified radiometers used for shortwave radiation into the following groups:

1. Reference standard pyrheliometers
    Angstrom compensation pyrheliometer (Sect. 5.3.1.1)
    Silver disc pyrheliometer (Sect. 5.3.1.2)

2. First class pyrheliometers
    Michelson bimetellic pyrheliometer (Sect. 5.3.1.3)
    Linke–Feussner pyrheliometer (Sect. 5.3.1.4)
    New Eppley pyrheliometer (Sect. 5.3.1.5)
    Yanishevsky thermoelectric pyrheliometer (Sect. 5.3.1.6)
3. Second class pyrheliometers
    Moll–Gorczynski (Sect. 5.3.1.7)
    Old Eppley pyrheliometer (before 1958)
4. First class pyranometers
    Selected thermopile pyranometers
5. Second class pyranometers
    Moll–Gorczynski pyranometer (Sect. 5.3.2.1)
    Eppley instruments (Sect. 5.3.2.2)
    Volochine thermopile pyranometer
    Dirmhirn–Sauberer pyranometer (Sect. 5.3.2.3)
    Yanishevsky thermoelectric pyranometer (Sect. 5.3.2.4)
    Spherical Bellani pyranometer (Section 5.3.2.5)
6. Third class pyranometers
    Robitzsch bimetallic pyranograph (Sect. 5.3.2.7)

5.3.1   Pyrheliometers

5.3.1.1   Angstrom Compensation Pyrheliometer

The compensation pyrheliometer is one of the best known and most reliable
instruments for measuring solar intensity (Fig. 5.1). Basically it consists of two

Figure 5.1. Angstrom compensation pyrheliometer (Courtesy of Eppley Laboratory).

manganin strips approximately $20 \times 2 \times 0.01$ mm, one of which is exposed to the sun's rays; the other manganin strip is heated electrically until it is at the same temperature as the strip exposed to solar radiation. Under steady state conditions, the energy used for heating is equal to the absorbed solar radiation. Thermocouples mounted on the back of the manganin strips are used to determine when both strips are at the same temperature. The flux density, E, of solar radiation is calculated from the formula $E = Ci^2$, where C is the instrument constant and i is the heating current in amperes. The instrument constant C can be calculated from knowledge of the resistance, the size of the manganin strip, the absorption of the blackening material, and the electrical heat equivalent. However, in practice C is usually determined by comparison with another standard instrument.

5.3.1.2   Silver Disc Pyrheliometer

The silver disc pyrheliometer was developed by Abbott of the Smithsonian Institute and has been used as a secondary standard to measure solar radiation for many years. It consists of a blackened silver disc positioned at the lower end of a tube with diaphragms to limit the hole aperture to $5.7°$. A mercury thermometer is used to measure the temperature of the disc. A triple shutter is used to alternately expose the silver disc to, and shade it from, solar radiation. The operational procedures must be followed very exactly because a consistent error of 1 s may result in an error of about 1% in the final result. These instruments have been found to be very good and are widely used for calibrating pyranometers.

5.3.1.3   Michelson Bimetallic Pyrheliometer

The sensing element consists of a bimetallic (usually constantan–invar) strip exposed to solar radiation. The deflection of the metal strip is observed by a low power microscope. Several models exist having angular apertures from approximately $5°$ to $25°$. The time required for full response is between 20 and 30 s.

5.3.1.4   Linke–Feussner Pyrheliometer

This pyrheliometer uses a Moll thermopile protected by a thick shell of copper made from a series of concentric conical milled rings screwed to one another (Fig. 5.2). This heavy mass is used to reduce the effect of wind and ambient temperature fluctuations. The thermopile consists of 18 junctions of manganin and constantan with a resistance of about $30\ \Omega$ and a sensitivity of $14\ \mu V\ W^{-1}\ m^2$. The aperture angle of the instrument is about $11°$. The older instruments had a temperature sensitivity of approximately -0.2% per $°C$. The newer instruments are temperature compensated. The

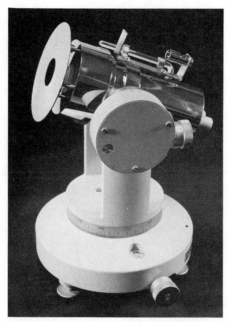

Figure 5.2. Linke–Feussner pyrheliometer (Courtesy of Kipp & Zonen).

time of response is approximately 10 s. A series of filters are available with this instrument to measure portions of the solar spectrum.

5.3.1.5   Eppley Normal Incidence Pyrheliometer

A temperature compensated 15-junction bismuth silver thermopile is mounted at the base of a brass tube constructed from a series of concentric rings yielding a conical angle of 5.7°. A quartz crystal is used to seal the tube filled with dry air. The instrument (Fig. 5.3) has a sensitivity of 8 μV W^{-1} m^2, an internal resistance of 200 Ω, and a response time of about 1 s. A filter wheel consisting of three standard meterological filters is available.

5.3.1.6   Yanishevsky Thermoelectric Pyrheliometer

This pyrheliometer utilizes a star-shaped thermopile having a sensitivity of 10 μV W^{-1} m^2, mounted in a tube with angular aperture of 10°. The 99% time constant is 25 to 30 s.

5.3.1.7   Moll–Gorczynski Pyrheliometer

This instrument incorporates a Moll-type thermopile consisting of between 10 and 80 junctions of manganin and constantan. The recording model has an angular aperture of about 8°. The thermopile was protected by a glass cover. It has a resistance of 50 to 60 Ω and a sensitivity of 36 μV W^{-1} m^2.

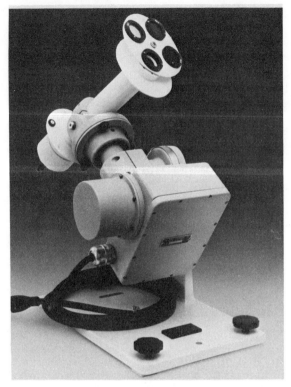

Figure 5.3. Eppley normal incidence pyrheliometer (Courtesy of Eppley Laboratory).

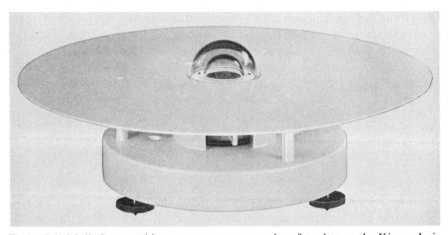

Figure 5.4. Moll–Gorczynski pyranometer, commonly referred to as the Kipp solarimeter (Courtesy of Kipp & Zonen).

5.3.2   Pyranometers

5.3.2.1   Moll–Gorczynski Pyranometer

This instrument contains a 14-junction manganin–constantan thermopile having a sensitivity 12 μV W^{-1} m^2, a resistance of approximately 10 Ω in the form of a 10 $\times$ 14 mm rectangle (Fig. 5.4). The thermopile is flush with the edge of its brass case with which the cold junctions are in contact. The blackened surface of the thermopile is covered by two concentric, optically ground, glass hemispheres 30 and 50 mm in diameter. The air space between the hemispheres is usually connected to a bottle of dessicant to prevent condensation on the inner surfaces. The 98% time constant is about 30 s. The thermopile is not compensated and has a temperature co-efficient of -0.2% per °C.

5.3.2.2   Eppley Instruments

5.3.2.2.1   Eppley (180° Pyrheliometer) Pyranometer

The thermal transducer of the Eppley pyrheliometer consists of a thermopile made from an alloy of 60% gold and 40% palladium wire against an alloy of 90% platinum and 10% rhodium. Alternate junctions are placed in thermal contact with concentric rings of silver. The receiving surface of the inner ring is black and that of the outer ring is white, representing the hot and cold junctions, respectively. The thermopile is hermetically sealed in a lamp bulb of soda lime glass. The glass bulb has uniform transmission from 0.35 to 2.5 μm and upper cutoff at approximately 4.5 μm. The bulb contains dry air to prevent condensation on the inner surface. Two models were available, one containing a 10-junction thermopile having a sensitivity of 3 μV W^{-1} m^2, a resistance of 35 Ω, and a time constant of 55 s. The other model contains 50 junctions, having a sensitivity of 11 μV W^{-1} m^2 and an internal resistance of 100 Ω. This instrument has been widely used in the United States. Care should be exercised when using this device for measuring albedo because of internal reflections and different sensitivities when the instrument is in the inverted position.

5.3.2.2.2   High-Precision Eppley Pyranometer

This instrument contains a 15-junction bismuth–silver thermopile with a thermistor temperature compensating circuit. The thermal transducer is blackened with optical black lacquer and is shielded from the weather by two removable concentric hemispheres. The instrument has a sensitivity of 10 μV W^{-1} m^2 and an internal resistance of 450 Ω. The 98% response time is approximately 30 s. The outer glass dome may be replaced with Schott optical glass filters, yellow GG14, orange OG1, RG2, or dark red RG8, representing lower cutoff values of 500, 530, 630, and 700 nm, respectively. The clear glass domes are transparent to wavelengths from 295 to 2800 nm.

Figure 5.5. Eppley precision spectral pyranometer (Courtesy of Eppley Laboratory).

5.3.2.2.3   Eppley Precision Spectral Pyranometer

This instrument is an improved smaller model of high precision Eppley pyranometer (Fig. 5.5). The major difference between the instruments is in the thermopile construction. This instrument utilizes a multijunction copper–constantan plated thermopile. The characteristics of the instrument include a sensitivity of 9 μV W^{-1} m^2 and impedance of 650 Ω. Sharp cutoff filters similar to the above are also available for this instrument.

5.3.2.2.4   Eppley Black and White Pyranometer

The detector consists of three white and three black wedges representing the cold and hot junctions of a electroplated copper–constantan thermopile (Fig. 5.6). The thermopile is protected by a single glass dome. It has a sensitivity of 11 μV W^{-1} m^2 and an impedance of 350 Ω. The time constant is 5 s.

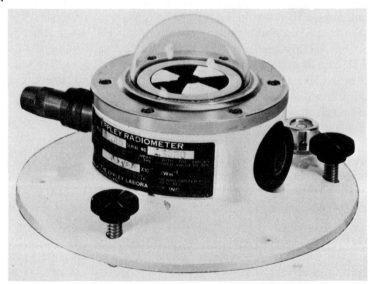

Figure 5.6. Eppley black and white pyranometer (Courtesy of Eppley Laboratory).

5.3.2.3   Dirmhirn–Sauberer Pyranometer

The Dirmhirn–Sauberer pyranometer consists of eight white and eight black sections radiating from a central point. Attached to these sections are 32 junctions of a thermopile. The black sections are warmed by radiant heat while the white sections remain at or near ambient temperature. The sensing surface is enclosed by a ground glass dome. The instrument has a sensitivity of approximately 2 μV W^{-1} m^2, an internal resistance of 5 Ω, and a time constant of about 30 s.

5.3.2.4   Yanishevsky Thermoelectric Pyranometer

This instrument utilizes a thermopile with black and white junctions. The sensing area is about 3 cm^2. It has a sensitivity of 10 μV W^{-1} m^2 and a time constant of about 6 s.

5.3.2.5   Bellani Pyranometer

The Bellani spherical pyranometer consists of a liquid-filled blackened copper sphere inside an evacuated glass globe. The inner sphere has a tube leading to a lower graduated condenser. Liquid, either alcohol or water, contained within the sphere is evaporated by solar radiation and condenses in the lower tube. The amount of liquid evaporated is a linear function of the shortwave energy incident on the inner sphere. The temperature coefficient of this device is approximately -0.2% per °C. The instrument is usually used to obtain daily totals or intervals of longer periods of time. Since its receptor is a sphere, measurements cannot be converted readily to flatplate radiometer equivalents.

5.3.2.6   Other Pyranometers

5.3.2.6.1   Middleton Solarimeter
The Middleton solarimeter contains a thermopile consisting of 82 copper–constantan junctions attached to a black sensor plate. This sensitive plate is enclosed by two concentric ground glass domes that protect the sensor plate against weather and heat convection distortions. This instrument has a sensitivity of 16 μV W^{-1} m^2 and internal resistance of 40 Ω. Similar sensitive elements have been constructed in special housings to be used for measuring albedo.

5.3.2.6.2   Davos Observatory
The sensing element of the Davos pyranometer has a 60-junction thermopile shielded by two concentric glass domes. This instrument has a sensitivity of 37 μV W^{-1} m^2 and a time constant of 3 s.

5.3.2.6.3   Spectran Pyranometer

This pyranometer has a single hemispherical dome, sensitivity of 4 μV W^{-1} m^2 and a time constant of 1 s. Its claimed accuracy is $\pm 2\%$.

5.3.2.7   Robitzsch Bimetallic Pyranograph

This instrument depends on the differential movement of two bimetallic strips (about 8.5 × 1.5 cm) arranged side by side. Both strips are exposed to the sun, one being blackened and the other painted white or highly polished to reflect solar energy. Several different configurations exist (Fig. 5.7), and some instruments utilize three strips. The differential movement of the bimetallic strips is recorded by a mechanical pen on a self-contained spring clock driven recorder. The sensitive portion is covered by a polished glass hemisphere approximately 130 mm in diameter. The sensitivity of this instrument depends on many factors, for example, the temperature of the

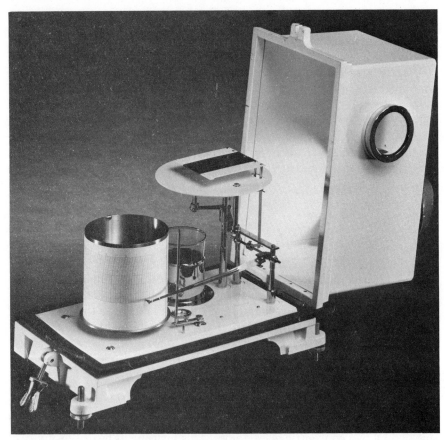

Figure 5.7. Bimetallic recording pyranometer (pyranograph) (Courtesy of Weather Measure Corporation).

case, the radiation intensity, the elevation, and azimuth of the sun. The time constant of this instrument is relatively great—about 2 min.

5.3.3   Pyrgeometer

Measurement of the longwave radiation portions of the sky or the earth at night can be accomplished by using any of the pyrheliometers without filters or windshields. Phyrheliometers that can be used for this purpose are the Angstrom (5.3.1.1), the Linke–Feussner (5.3.1.4), and the Abbot silver disc (5.3.1.2). Since these instruments are designed with a small angle of view, it is necessary to scan sections of the hemisphere and correct for cosine response to obtain hemispherical radiation. At night a total hemispherical radiometer will provide values of longwave sky or terrestrial radiation.

At present, the only single instrument capable of measuring longwave radiation from a hemisphere in the presence of sunlight is the Eppley precision infrared radiometer. This instrument is similar to 5.3.2.2.3 with the addition of a special interference filter that is effectively opaque to radiation of less than 4 μm and a circuit to compensate for outgoing longwave radiation. This instrument has a sensitivity of 9 μV W^{-1} m^2, an impedance of 650 Ω, and time constant of 1 s.

During the daytime, global solar radiation can be subtracted from total hemispherical radiation to obtain longwave hemispherical radiation.

5.3.4   Total Hemispherical Radiometers

Total hemispherical radiation includes both longwave and shortwave radiation from a hemisphere. Instruments used are divided into two types: those that are ventilated and those that are shielded to reduce the convective heat transfer from the sensing element.

5.3.4.1   Gier and Dunkle Radiometer

The total hemispherical radiometer was described by Gier and Dunkle (1951). It contains a thermopile consisting of 300 silver–constantan junctions embedded in a plate approximately 114 mm^2 (Fig. 5.8). Both surfaces of the plate are ventilated by a blower to equalize the thermal convection term. The upper surface is blackened. The lower surface of the transducer is polished aluminum and is protected from upward radiation by another sheet of aluminum blackened on the top side and polished on the bottom side. Thus, the lower portion of the transducer is at an ambient temperature while the temperature of the upper portion depends on incident radiation. This instrument has a sensitivity of approximately 14 μV W^{-1} m^2 and a time response of 10 s.

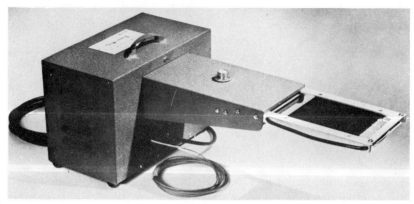

Figure 5.8. Gier and Dunkle-type total hemispherical radiometer (Courtesy of Science Associates).

5.3.4.2 Schulze Radiometer

The Schulze radiometer (Schulze, 1961) is basically two total hemispherical radiometers placed back to back (Fig. 5.9). The thermopiles are made of constantan and silver and have a resistance of about 500 Ω. The sensitive portion is blackened with optical black lacquer and shielded from the effects of weather with a thin plastic hemispherical shield constructed from polyethylene (lupolen-*h*). In operation the instrument may be used to measure

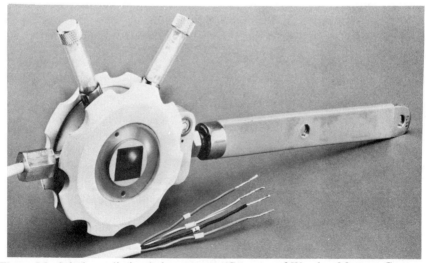

Figure 5.9. Schulze radiation balance meter (Courtesy of Weather Measure Corporation).

total hemispherical radiation, either downward or upward, or by suitable connection it can be used to determine net radiation directly. Since no ventilation is required, the instrument may be operated without an external power source.

5.3.4.3   Davos Observatory

The sensing element is similar to that used in the pyranometer (see 5.3.2.6.2). Its element is protected by a polyethylene hemisphere. A configuration consisting of two opposing pyranometers and two opposing total hemispherical radiometers is available that can be used to measure the four components of radiation.

5.3.4.4   Economical Net Radiometer

The economical net radiometer has undergone modifications. The latest version is described by Tanner et al. (1969). Basically it consists of two blackened sensing elements separated by a layer of insulating material or air space. These elements are shielded from the effects of weather by two sheets of polyethylene film. The temperature of the sensing element may be determined by use of mercury thermometers, dial thermometers, or electrical devices such as thermocouples and thermistors. Total hemispherical radiation may be calculated from either side, or when combined together may be used to calculate net radiation. Because flat polyethylene windshields are used, the cosine response of this instrument is rather poor to zenith angles greater than about 65°. The instrument has a time constant of 15 min and cannot be considered as an all-weather type instrument. However, it is very convenient to use to make spot measurements of radiation because it does not require any source of power.

5.3.4.5   Modified Shielded Net Radiometers

Shielded net radiometers (5.3.5) have been modified by placing an adaptor on either side. The black body radiation of the radiation adaptor is added to the measured net radiation to obtain the total hemispherical radiation.

5.3.5   Net Radiometers

As with total hemispherical radiometers, net radiometers are divided into ventilated and shielded types. The more commonly used types are described here. The Schulze Radiation Balance Meter is described in Sect. 5.3.4.2, and the Davos Observatory Net Radiometer is described in Sects. 5.3.2.6.3 and 5.3.4.3.

5.3.5.1   Gier and Dunkle Ventilated Net Radiometer

This instrument is basically the same as the Gier and Dunkle total hemispherical radiometer described in Sect. 5.3.4.1. The difference lies in the fact that the transducer is blackened on both surfaces and the lower aluminum plate is eliminated. Two models are available, the standard and the temperature compensated model. The standard model has a sensitivity of 14 μV W^{-1} m^2, a 95% response time of 12 s, and an air flow of 11 m s^{-1}. The temperature compensated model has similar specifications with the exception that the sensitivity is reduced to 2 μV W^{-1} m^2.

5.3.5.2   Suomi Improved Net Radiometer

The thermal transducer (Suomi, 1954) consists of 120 copper–constantan junctions wound around a microscope slide yielding a sensitivity of approximately 3 μV W^{-1} m^2. The main difference between this instrument and the Gier and Dunkle net radiometer is the relatively high conductance of the thermal transducer, a mechanism for balancing the ventilation, and an increased ventilation rate of approximately 22 m s^{-1}. This instrument is not available commercially.

5.3.5.3   Funk Net Radiometer

Two models of the Funk radiometer exist. The larger radiometer (Funk, 1959) contains a thermal transducer having 250 thermal junctions bonded by two blackened plates (Fig. 5.10). Hermispherically formed windshields made from polyethylene are used to reduce the thermal convection term. This instrument has a sensitivity of 57 μV W^{-1} m^2, internal resistance of 180 Ω, and is approximately 16 cm in diameter including the heating ring. The miniature version (Funk, 1962) contains 45 thermal junctions and has a sensitivity of 8 μV W^{-1} m^2, with an internal resistance of approximately 23 Ω. The sensing head of the miniature version is 14 mm in diameter.

Figure 5.10. Funk-type net radiometer (Courtesy of Science Associates).

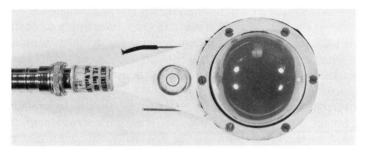

Figure 5.11. Fritschen-type net radiometer.

5.3.5.4 Fritschen Net Radiometer

The temperature compensated net radiometer (Fritschen, 1965) contains a 22-junction manganin–constantan thermopile in circuit with a thermistor temperature compensator (Fig. 5.11). The sensitive surface is blackened and is enclosed with hemispherically shaped polyethylene windshields 5 cm in diameter. This instrument has a sensitivity of 5 μV W^{-1} m^2, internal resistance of 200 Ω, and a time constant of approximately 12 s.

5.3.6 Diffuse Radiation

Separation of direct beam and diffuse radiation requires the use of either a pyrheliometer and pyranometer or a pyranometer with and without a shadow band or shading disc. The shading disc requires an equatorial mount similar in principle to that used with pyrheliometer to keep the disc directly between the sun and the sensing element of the pyranometer. In addition, the disc should be large enough to shade the glass domes of the pyranometer, thereby preventing internal reflections.

Shadow bands are easier to use than shading discs. Design criteria have been presented (Blackwell, 1954; Durmmond, 1956; Horowitz, 1969). When shadow bands are used, the measured diffuse radiation should be corrected for the portion of the sky obliterated by the band. Latimer (1971) gives corrections for 0° to 90° N latitude.

5.4 Site Requirements

The site should be free of any obstructions above the level of the sensing element for pyranometers and also below the level of the sensing element for albedo and net radiation measurements. This is especially true for the angles from ENE through S to WNW in the northern hemisphere. Care should be taken to ensure that towers, mast, or telephone poles do not cast a shadow on the sensing element during the day. If possible, the site should be chosen so that these elements are below 5° elevation from horizontal. This is especially true when instruments are located near light colored buildings with highly reflective surfaces such as aluminum. Generally, the best site for locating a pyranometer is the roof of a building. The same rules apply to obstructions above the plane of the sensing element when the instrument is mounted on the roof with respect to firewalls, ventilator shafts, or such other items. However, the location should be readily accessible for inspection and maintenance.

An inverted pyranometer is used in surface albedo determinations. If one instrument is used for incoming and reflection measurement, it should be mounted so that it may be readily inverted from its normal position. Care should be taken to ensure that, in the downward facing position, the receiver surface remains horizontal. The mounting should introduce a minimum of obstructions to radiation. A series of readings should be taken alternately exposing the pyranometer to radiation from the earth's surface and the sky. The calibration factor of the instrument should be checked in the inverted position because the calibration factor of some instruments change when the instrument is inverted. If an Eppley 180° pyrheliometer is used to determine the albedo, a shield should be placed around the bulb to eliminate internal reflections within the bulb. Additional shielding is required to prevent the base of most inverted pyranometers from overheating.

Net radiometers should be located over the surface in question if meaningful results are to be obtained. The mounting apparatus for supporting the net radiometer should be chosen as to present a minimum of obstruction to radiant fluxes by casting shadows. Over surfaces such as grass or alfalfa the height of the instrument is not very critical. However, with row crops spaced at 1 m intervals, the instrument should be located at least 1 m above the surface to be representative of a large area. For widely scattered type vegetation, a higher location is desirable.

5.5   Calibration

5.5.1   Pyranometer Calibration

Several methods of calibrating pyranometers are available. The preferred method is to compare the pyranometer in question with a primary or a secondary pyrheliometer using the sun as the source of radiation. Measurements are made on a clear day, preferably at solar noon. The pyranometer is periodically shaded with a disc mounted on a long slender arm so as just to obscure the solid angle of the sun. The difference in the signal between the shaded and unshaded readings represents the energy from the direct solar beam of the sun. This may be compared with the readings from a standard pyrheliometer after the mean solar elevation has been taken into account. It is important that the level of the instruments used be carefully checked. A small difference in level can cause rather serious errors in calibration factors.

The pyranometer can be checked by comparing it against another pyranometer for a 24-hour period. This is best done with clear skies and several times during the year if either instrument has a temperature coefficient.

Pyranometers are also calibrated in artificial light chambers called integrating spheres where the radiation field can be stabilized and ambient temperatures can be varied (MacDonald and Foster, 1954). This method is far more reliable and easier to perform than the outdoor technique. Calibrations are repeatable to $\pm 1 \%$.

5.5.2   Net Radiometer Calibration

Net radiometers may also be calibrated by the shading technique similar to that described for the pyranometer. Net radiometers require an additional check of symmetry. This is especially important for the ventilated type. If the difference between the top and bottom readings exceed 2 or 3%, the instrument should be corrected.

A net radiometer may be checked while in use by comparing it to the output of another similar type net radiometer. The second net radiometer should be used only for this purpose and should be stored in a place where it is unlikely to be damaged.

Several laboratory techniques for calibrating net radiometers have been described. Johnson (1956) described a chamber which utilized a tungsten lamp as a radiation source. The radiation entered the blackened test chamber through a small aperture in the top. The longwave component of the radiation was filtered with four sheets of window glass, two located above and two located below the test area. The radiant flux density from the lamp was measured with a pyranometer. The difference in signal from a net radiometer, exposed to the radiation from the lamp and with the aperture closed, was equated to the radiant flux density previously measured. The above chamber design was modified (Fritschen, 1963) by the omission of the glass filters and the inclusion of independent temperature controlled liners on the top and bottom half of the chamber to enable longwave calibrations.

Funk (1961) concluded that longwave calibration procedures that enclosed each half of a net radiometer in separate radiation cavities should be avoided because differential unwanted sensible heat flux from the shields would increase the apparent sensitivity of the instrument.

A non-steady-state longwave calibration procedure was described by Idso (1971). The net radiometer was located above a heated or cooled blackened plate in a constant temperature room. The slope of the signal from the radiometer versus the radiant flux from the plate as it is heated or cooled constituted the calibration factor.

5.6   Photometry

Photometry, unlike radiometry, deals only with the visible portion of radiant energy and its effect upon the eye.

5.6.1   Photometric Measures

The measurement of light, like radiometry, requires the measurement or the calculation of energy at some point in space. If this radiant energy is in the visible spectrum, the energy has the ability to provide the sensation of sight or make a surface visible. Consequently it is called luminous energy. The basic unit of measurement adopted by the International Committe of Weights and Measures in 1937 is called the *candela* (cd). It is defined as one-sixtieth of the luminous intensity of one square centimeter of a black body radiator at the temperature of solidification of platinum.

The definitions of photometric measures parallel those of radiometric measures with the word luminous used in place of radiant (see Table 5.3). Radiometric measures were discussed in detail in Section 2.1.2.4. Briefly, the definition of the *flux* of the source is the rate at which energy is emitted. The *flux density* at a distance r, m, from the center of the source (irradiance or illuminance) is the total flux divided by the surface area over which it is spread. The *intensity* of a source is the total flux divided by the total solid angle through which it is distributed. *Emittance* is the flux density at the surface

Table 5.3. A comparison of radiometric and photometric measures.

Radiometric	Photometric
Flux (Φ), Watt (W)	Flux (F), lumen (lm)
Flux density (E)	Flux density (E_i)
$E = \Phi/A$, W m^{-2}	$E_i = F/A$, lm m^{-2} or lux (lx)
Irradiance	Illuminance
Intensity (I),	Intensity (I_i),
$I = \Phi/\omega$, W sr^{-1}	$I_i = F/\omega$, lm sr^{-1} = candela (cd)
Radiant intensity	Luminous density
Emittance (M)	Emittance (M_i)
$M = \Phi/A$, W m^{-2}	$M_i = F/A$, lm m^{-2} or cd sr m^{-2}
Radiant emittance	Luminous emittance
Angular emittance (L)	Angular emittance (B)
$L = \Phi/\omega A$, W (m^2 sr)$^{-2}$	$B = F/\omega A$, lm (m^2 sr)$^{-1}$
Radiance	Luminance

of the source while *angular emittance* (radiance or luminance) is the intensity per unit area of the source.

5.6.2   Photopic Vision

Photopic vision concerns radiant energy of wavelengths 0.38–0.77 μm acting on the sensitive nerves within the retina of the eye. Two types of nerve cells are located in the retina—cones and rods. When the radiant energy levels are high the cones are activated and the resulting vision is called *photopic vision*. When the radiant energy levels are low and rods are activated the vision is *scotopic*. We will limit our discussion to photopic vision. Color and brightness are the two sensations of sight. The color depends only on the spectral composition of the radiant energy.

5.6.3   Luminous Efficiency

The instruments described to measure radiation generally had uniform sensitivity across a wide range of wavelengths. The eye is a very selective receiver. It is more sensitive to visible radiation having peak sensitivity at 0.5505 μm. The relative sensitivity at other wavelengths is shown in Fig. 5.12 while exact values are given in Table 5.4. The ratio of spectral radiant flux at 0.5505 μm required to produce a specific brightness sensation to the flux at wavelength, λ, required to produce the same brightness sensation is called *relative luminous efficiency*, V_λ. The values given in Table 5.4 serve to define a "standard observer."

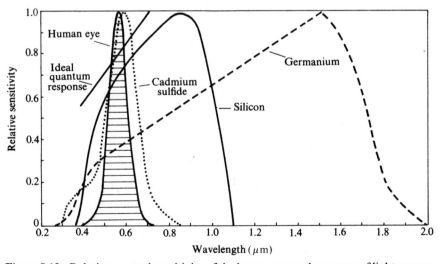

Figure 5.12. Relative spectral sensitivity of the human eye and response of light sensors.

Table 5.4. Spectral luminous efficiency of the standard observer
relative to 0.55 μm.

Wavelength (μm)	V_λ
0.40	0.000 4
0.45	0.038
0.50	0.323
0.55	0.995
0.550 5	1.000
0.60	0.631
6.65	0.107
0.70	0.004 1
0.75	0.000 1

5.6.4 Relation Between Photometric and Radiometric Measures

The reflection of radiant flux by a surface in the visible portion of the spectrum is called luminous flux and causes the irradiated surface to be visible. Thus, the luminous flux, F, is directly related to the radiant flux, Φ, and the relative luminous efficiency V_λ by $F(\lambda) = K_m V_\lambda \Phi(\lambda)$ where K_m is a constant independent of wavelength, 686.5 lm W^{-1}. The product $K_m V_\lambda$ is referred to as luminous efficiency of monochromatic radiation. Similarily, the relation between other photometric and radiometric measures can be defined:

$$I_i = I K_m V_\lambda$$
$$M_i = M K_m V_\lambda$$
$$E_i = E K_m V_\lambda$$
$$B = L K_m V_\lambda$$

The above relations can be extended to band portions of the spectrum. For example, photosynthesis is activated by radiant energy between 0.4 and 0.7 μm. This band is referred to as *photosynthetically active radiation* (PAR). The median of this band is 0.51 μm.

5.6.5 Light Sensors

Light sensors utilize photocells made from cadmium sulfide, germanium, or silicon and some combination of filters. The selection of a specific material is dependent on its spectral sensitivity and stability. These are compared with the sensitivity of the eye in Fig. 5.12. The sensitivity of the cadmium sulfide cell closely matches the sensitivity of the eye, consequently it is especially useful in applications where human vision is a factor. Examples are automatic street lights, yard light control, and automatic iris control in cameras. Germanium responds relatively fast to changes in light and is useful for high frequency switching. Silicon cells are generally useful for direct conversion into electrical power.

Norman *et al.* (1969) described the filtering of two silicon cells to produce a quantum response sensor from 0.4 to 0.7 μm (PAR). The response was achieved by obtaining the difference signal from a cell sensing infrared and a cell sensing both visible and infrared radiation. Both cells were placed under heat absorbing glass and were cosine corrected with diffusing plastic. A commercial version using silicon photodiode is produced by Lambda Instruments.

Bibliography

Coulson, Kinsell L. (1975). Solar and Terrestrial Radiation. Academic, New York. 322 pp.

Reifsnyder, W. E., and H. W. Lull (1965). Radiant energy in relation to forest. *U.S. Dep. Agric. Tech. Bull.* **1344**.

Robinson, N. (1966). Solar Radiation. Elsevier, New York. 347 pp.

World Meteorological Organization (1971). Guide to Meteorological Instrument and Observing Practices. World Meteorological Organization, Geneva, Switzerland.

I.G.Y. (1958). Instrumentation Manual 6, Radiation Instruments and Measurements. Pergamon, New York, N.Y.

Literature Cited

Alrich, L. B., H. Wexler, S. Fritz, I. F. Hand, A. Court and W. P. Mellen, 1947. Comments by Readers. Sci. 106:225.

Blackwell, M. J. (1954). Five Years Continuous Recording of Daylight Illumination at Kew Observatory. Met. Res. Ctr. Air Ministry, London M.R.P. No. 831.

Drummond, A. J. (1956). Notes on the measurement of natural illumination I. Some characteristics of illumination recorders. *Arch. Meteorol. Geophys. Bioklimatol. Ser.* B 7:437.

Fritschen, L. J. (1963). Construction and evaluation of a miniature net radiometer. *J. Appl. Meteorol.* **2**:165–172.

Fritschen, L. J. (1965). Miniature net radiometer improvements. *J. Appl. Meteorol.* **4**:528–532.

Funk, J. P. (1959). Improved polythene-shield net radiometer. *J. Sci. Instrum.* **36**:267–270.

Funk, J. P. (1961). A note on the longwave calibration of convectively shielded net radiometers. *Arch. Meteorol. Geophys. Bioklimatol. Ser.* B 11:70–74.

Funk, J. P. (1962). A net radiometer designed for optimum sensitivity and a ribbon thermopile used in a miniature version. *J. Geophys. Res.* **67**:2753–2760.

Gier, J. T. and R. V. Dunkle (1951). Total hemispherical radiometers. AIEE Trans. **70**:1–7.

Horowitz, J. L. (1969). An easily constructed shadow-band for separating direct and diffuse solar radiation. *Sol. Energy* **12**:543–545.

Idso, Sherwood, B. (1971). A simple technique for the calibration of long-wave radiation probes. *Agric. Meteorol.* **8**:235–243.

Johnson, D. S. (1956). Progress Report on Radiometer Test. U.S. Weather Bureau unpublished mimeogr. rpt. December, 20 pp.

Latimer, J. R. (1971). Radiation measurement. International Field Year for the Great Lakes. Tech. Manual Ser. 2. Canadian National Committee for the International Hydrological Decade. No. 8 Building, Carling Avenue, Ottawa, Canada. p. 52.

MacDonald, T. H. and N. B. Foster (1954). Pyrheliometer calibration program of the U.S. Weather Bureau. *Mon. Weather Rev.* **82**:219–227.

Norman, J. M., C. B. Tanner, and G. W. Thurtel (1969). Photosynthetic light for measurements in plant canopies. *Agron. J.* **61**:840–843.

Schulze, R. (1961). Über die Verwendung von Polyäthylen für Strahlungsmessungen. *Arch. Meteorol. Geophys. Bioklimat.* Ser. B **11**:211–223.

Suomi, V. E., M. Franssila, and N. F. Islitzer (1954). An improved net-radiation instrument. *J. Meteorol.* **11**:276–282.

Tanner, C. B., C. A. Federer, T. A. Black, and J. B. Swan (1969). Economical Radiometer, Theory, Performance, and Construction. Research Report 40. College of Agriculture and Life Sciences, Univ. of Wisconsin, Madison, Wisc. p. 86.

Chapter 6

Humidity and Moisture

6.1 Fundamental Concepts and Definitions

Knowledge of the basic definitions in common use is essential for an understanding of the principles of moisture measurement.

The pressure, volume, and temperature relationships for an ideal gas are

$$PV = nRT, \tag{6.1}$$

where P is the pressure of the gas, V is volume of gas, n is the number of moles of the gas, R is the universal gas constant, and T is absolute temperature. The number of moles is related to the mass of the gas, m, and its molecular weight, M, by

$$n = \frac{m}{M}, \tag{6.2}$$

while the density, ρ, is related to mass and volume by

$$\rho = \frac{m}{V}. \tag{6.3}$$

The total pressure of moist air, P, consists of the partial pressure of dry air, P_a, plus that of the water vapor, e, according to Dalton's law of partial pressures,

$$P = P_a + e. \tag{6.4}$$

The *absolute humidity*, ρ_v, is the mass of water vapor per unit volume of air,

$$\rho_v = \frac{m_v}{V}, \tag{6.5}$$

where the subscript v represents vapor. A useful form of the previous equation is the conversion of vapor density to pressure by $\rho_v = 2.17 \times 10^{-3}e/T$ where ρ_v is in $kg\,m^{-3}$ and e is in Pa.

The *mixing ratio*, r, is the ratio of the mass of water vapor to the mass of *dry* air,

$$r = \frac{m_v}{m_a}. \tag{6.6}$$

The mixing ratio may also be expressed as

$$r = \left(\frac{M_v}{M_a}\right) \frac{e}{(P - e)}, \tag{6.7}$$

which reduces to

$$r = 0.622 \frac{e}{(P - e)}, \tag{6.8}$$

by substituting the ratio of the molecular weights of water vapor and dry air. Since $e \ll P$, $r \simeq 0.622e/P$.

The ratio of the mass of water vapor to the mass of moist air is called the *specific humidity*, q,

$$q = \frac{m_v}{(m_v + m_a)}. \tag{6.9}$$

But q also may be expressed in terms of molecular weight and pressure as

$$q = \frac{M_v e}{(M_v e + M_a P_a)}. \tag{6.10}$$

Dividing numerator and denominator by M_a and substituting $(P - e)$ for P_a yields

$$q = \left(\frac{M_v}{M_a}\right) \frac{e}{\{P - [1 - (M_v/M_a)]e\}}. \tag{6.11}$$

Since $1 - (M_v/M_a) = 0.378$, Eq. (6.11) becomes

$$q = 0.622 \frac{e}{(P - 0.378e)}. \tag{6.12}$$

The concept of saturation is required in the definition of relative humidity, dew-point temperature, and wet-bulb temperature. *Saturation* is the state in which moist air coexists in neutral equilibrium with an associated condensed phase of either water or ice at a given temperature and pressure. The surface of separation must be a plane, or surface tension effects will shift the equilibrium. The *saturation vapor pressure*, e_w, of pure aqueous vapor with respect to water is the pressure of the vapor when in thermal equilibrium with a plane of pure water at the same temperature and pressure. In the meteorological range of temperature and pressure, the saturation vapor pressure of moist air represents the maximum amount of water vapor that can be held by air of a given temperature and pressure. At constant pressure, the saturation vapor pressure increases with temperature, nearly doubling for each 10°C increase. In Fig. 6.1 the uppermost curved line represents the saturation vapor pressure at 101.3 kPa pressure. The saturation vapor pressure of water at 0°C is 0.610 78 kPa. Saturation vapor pressure for other temperatures computed with the Goff and Gratch (1946) formula are given in Table 6.1.

The *saturation deficit* or the *vapor pressure deficit*, VPD, is the difference between the saturation and actual vapor pressure at the same temperature and pressure. For example, at an air temperature of 29°C and vapor pressure of 2.005 kPa, the saturation vapor pressure is 4.005 kPa. Therefore the saturation deficit is 2.0 kPa.

The *relative humidity* of air, U, is the ratio in percent of water vapor of moist air relative to the saturation vapor pressure at the same temperature and pressure,

$$U = 100\left(\frac{e}{e_w}\right). \tag{6.13}$$

Relative humidity of 50%, labeled on the right axis of Fig. 6.1, represents the condition where the atmospheric vapor pressure is equal to one half of the saturation vapor pressure at that temperature. Both saturation vapor pressure and relative humidity may be defined with respect to a plane surface of ice. The saturation vapor pressure of ice at 0°C is practically equal to that over water at 0°C.

The *dew point*, T_d, is the temperature at which saturation will occur if moist air is cooled at constant pressure. This is also the condition in which the relative humidity is 100% and condensation occurs. To find the dew-point temperature of air at 50% relative humidity and 29°C temperature locate the intersection of 50% and 29°C in Fig. 6.1 and follow the 2.0 kPa line horizontally to the left until the 100% relative humidity line is reached. The isotherm intersecting this point, 17.6°C, is the dew-point temperature.

The *wet-bulb* temperature, T_w, of moist air at a given pressure and air temperature is the temperature attained when the moist air is brought adiabatically to saturation by evaporation of water into the moist air. The wet-bulb temperature for air of 50% U and 29°C air (Fig. 6.1) is found by moving from the 50% and 29°C intersection upward to the left parallel to the diagonal lines until the 100% U line is reached. The isotherm intersecting this point, 21.2°C, is the wet-bulb temperature.

Using Fig. 6.1, the saturation vapor pressure, vapor pressure, vapor pressure deficit, wet-bulb temperature, dew-point temperature, relative humidity, or air temperature may be found if any two of the above are known.

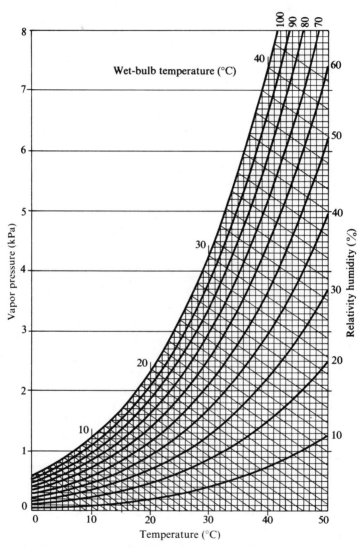

Figure 6.1. Temperature, vapor pressure, and relative humidity diagram.

Table 6.1. Saturation vapor pressure in Pa over ice when $T_w < 0$ C and over water when $T_w > 0°C$.

°C	0.0	0.1	0.2	0.3	0.4	0.5	0.6	0.7	0.8	0.9
−49	04.4	04.4	04.3	04.3	04.2	04.2	04.1	04.1	04.0	04.0
−48	05.0	05.0	04.9	04.8	04.8	04.7	04.7	04.6	04.6	04.5
−47	05.7	05.6	05.5	05.5	05.4	05.3	05.3	05.2	05.1	05.1
−46	06.4	06.3	06.2	06.2	06.1	06.0	06.0	05.9	05.8	05.7
−45	07.2	07.1	07.0	06.9	06.9	06.8	06.7	06.6	06.5	06.5
−44	08.1	08.0	07.9	07.8	07.7	07.6	07.5	07.5	07.4	07.3
−43	09.1	09.0	08.9	08.8	08.7	08.6	08.5	08.4	08.3	08.2
−42	10.2	10.1	10.0	09.9	09.8	09.6	09.5	09.4	09.3	09.2
−41	11.5	11.3	11.2	11.1	10.9	10.8	10.7	10.6	10.5	10.3
−40	12.8	12.7	12.5	12.4	12.3	12.1	12.0	11.9	11.7	11.6
−39	14.4	14.2	14.0	13.9	13.7	13.6	13.4	13.3	13.1	13.0
−38	16.1	15.9	15.7	15.5	15.4	15.2	15.0	14.9	14.7	14.5
−37	17.9	17.7	17.6	17.4	17.2	17.0	16.8	16.6	16.4	16.2
−36	20.0	19.8	19.6	19.4	19.2	19.0	18.7	18.5	18.3	18.1
−35	22.3	22.1	21.8	21.6	21.4	21.1	20.9	20.7	20.5	20.2
−34	24.9	24.6	24.3	24.1	23.8	23.6	23.3	23.1	22.8	22.6
−33	27.7	27.4	27.1	26.8	26.5	26.2	26.0	25.7	25.4	25.1
−32	30.8	30.5	30.1	29.8	29.5	29.2	28.9	28.6	28.3	28.0
−31	34.2	33.9	33.5	33.2	32.8	32.5	32.1	31.8	31.4	31.1

Table 6.1. (*cont.*)

°C	0.0	0.1	0.2	0.3	0.4	0.5	0.6	0.7	0.8	0.9
−30	38.0	37.6	37.2	36.8	36.4	36.1	35.7	35.3	34.9	34.6
−29	42.1	41.7	41.3	40.8	40.4	40.0	39.6	39.2	38.8	38.4
−28	46.7	46.2	45.7	45.3	44.8	44.4	43.9	43.5	43.0	42.6
−27	51.7	51.2	50.7	50.1	49.6	49.1	48.6	48.1	47.7	47.2
−26	57.2	56.6	56.1	55.5	54.9	54.4	53.8	53.3	52.8	52.2
−25	63.2	62.6	62.0	61.4	60.8	60.1	59.5	59.0	58.4	57.8
−24	69.8	69.2	68.5	67.8	67.1	66.5	65.8	65.2	64.5	63.9
−23	77.1	76.3	75.6	74.9	74.1	73.4	72.7	72.0	71.2	70.5
−22	85.0	84.2	83.4	82.6	81.8	81.0	80.2	79.4	78.6	77.9
−21	93.7	92.8	91.9	91.0	90.1	89.3	88.4	87.5	86.7	85.9
−20	103.2	102.2	101.2	100.2	99.3	98.3	97.4	96.5	95.5	94.6
−19	113.5	112.4	111.4	110.3	109.3	108.2	107.2	106.2	105.2	104.2
−18	124.8	123.6	122.5	121.3	120.2	119.0	117.9	116.8	115.7	114.6
−17	137.1	135.9	134.6	133.3	132.1	130.8	129.6	128.4	127.2	126.0
−16	150.6	149.2	147.8	146.4	145.0	143.7	142.4	141.0	139.7	138.4
−15	165.2	163.6	162.1	160.7	159.2	157.7	156.3	154.8	153.4	152.0
−14	181.1	179.4	177.8	176.2	174.5	173.0	171.4	169.8	168.2	166.7
−13	198.4	196.6	194.8	193.0	191.3	189.5	187.8	186.1	184.4	182.7

-12	217.2	215.2	213.3	211.4	209.4	207.6	205.7	203.8	202.0	200.2
-11	237.6	235.4	233.3	231.3	229.2	227.1	225.1	223.1	221.1	219.1
-10	259.7	257.4	255.1	252.9	250.6	248.4	246.2	244.0	241.8	239.7
-9	283.7	281.2	278.8	276.3	273.9	271.5	269.1	266.7	264.4	262.0
-8	309.7	307.0	304.4	301.7	299.1	296.5	293.9	291.3	288.8	286.2
-7	337.9	335.0	332.1	329.2	326.4	323.6	320.8	318.0	315.2	312.5
-6	368.5	365.3	362.2	359.0	356.0	352.9	349.9	346.8	343.9	340.9
-5	401.5	398.0	394.7	391.3	387.9	384.6	381.3	378.1	374.9	371.6
-4	437.2	433.5	429.8	426.2	422.5	419.0	415.4	411.9	408.4	404.9
-3	475.7	471.7	467.8	463.8	459.9	456.1	452.2	448.4	444.6	440.9
-2	517.3	513.0	508.8	504.5	500.3	496.1	492.0	487.9	483.8	479.7
-1	562.3	557.6	553.0	548.4	543.9	539.4	534.9	530.5	526.1	521.7
0	610.8	615.2	619.7	624.2	628.8	633.3	637.9	642.6	647.2	651.9
1	656.6	661.4	666.1	670.9	675.8	680.7	685.6	690.5	695.5	700.4
2	705.5	710.5	715.6	720.7	725.7	731.1	736.3	741.6	746.9	752.2
3	757.5	762.9	768.3	773.8	779.3	784.8	790.4	796.0	801.6	807.2
4	812.9	818.7	824.5	830.3	836.1	842.0	847.9	853.8	859.8	865.9
5	871.9	878.0	884.2	890.3	896.6	902.8	909.1	915.4	921.8	928.2
6	934.6	941.1	947.7	954.2	960.8	967.5	974.2	980.9	987.7	994.5
7	1001	1008	1015	1022	1029	1036	1043	1050	1058	1065
8	1072	1079	1087	1094	1102	1109	1117	1124	1132	1140
9	1147	1155	1163	1171	1179	1187	1195	1203	1211	1219
10	1227	1235	1244	1252	1261	1269	1277	1286	1295	1303
11	1312	1321	1329	1338	1347	1356	1365	1374	1383	1393
12	1402	1411	1420	1430	1439	1449	1458	1468	1477	1487
13	1497	1507	1517	1527	1537	1547	1557	1567	1577	1587

Table 6.1. (*cont.*)

°C	0.0	0.1	0.2	0.3	0.4	0.5	0.6	0.7	0.8	0.9
14	1 598	1 608	1 619	1 629	1 640	1 650	1 661	1 672	1 683	1 693
15	1 704	1 715	1 726	1 738	1 749	1 760	1 771	1 783	1 794	1 806
16	1 817	1 829	1 841	1 852	1 864	1 876	1 888	1 900	1 912	1 924
17	1 937	1 949	1 961	1 974	1 986	1 999	2 012	2 024	2 037	2 050
18	2 063	2 076	2 089	2 102	2 115	2 129	2 142	2 156	2 169	2 183
19	2 196	2 210	2 224	2 238	2 252	2 266	2 280	2 294	2 308	2 323
20	2 337	2 352	2 366	2 381	2 396	2 411	2 426	2 441	2 456	2 471
21	2 486	2 501	2 517	2 532	2 548	2 563	2 579	2 595	2 611	2 627
22	2 643	2 659	2 675	2 692	2 708	2 725	2 741	2 758	2 775	2 792
23	2 809	2 826	2 843	2 860	2 877	2 895	2 912	2 930	2 947	2 965
24	2 983	3 001	3 019	3 037	3 056	3 074	3 092	3 111	3 130	3 148
25	3 167	3 186	3 205	3 224	3 243	3 263	3 282	3 302	3 321	3 341
26	3 361	3 381	3 401	3 421	3 441	3 462	3 482	3 503	3 523	3 544
27	3 565	3 586	3 607	3 628	3 649	3 671	3 692	3 714	3 736	3 758
28	3 780	3 802	3 824	3 846	3 869	3 891	3 914	3 937	3 959	3 982
29	4 005	4 029	4 052	4 076	4 099	4 123	4 147	4 171	4 195	4 219
30	4 243	4 267	4 292	4 317	4 341	4 366	4 391	4 417	4 442	4 467
31	4 493	4 518	4 544	4 570	4 596	4 622	4 649	4 675	4 702	4 728
32	4 755	4 782	4 809	4 836	4 864	4 891	4 919	4 947	4 974	5 003
33	5 031	5 059	5 087	5 116	5 145	5 174	5 203	5 232	5 261	5 290
34	5 320	5 350	5 380	5 410	5 440	5 470	5 500	5 531	5 562	5 593

35	5 624	5 655	5 686	5 718	5 749	5 781	5 813	5 845	5 877	5 910
36	5 942	5 975	6 008	6 041	6 074	6 107	6 141	6 174	6 208	6 242
37	6 276	6 311	6 345	6 380	6 414	6 449	6 484	6 520	6 555	6 591
38	6 626	6 662	6 698	6 735	6 771	6 808	6 845	6 881	6 919	6 956
39	6 993	7 031	7 069	7 107	7 145	7 183	7 222	7 261	7 299	7 339
40	7 378	7 417	7 457	7 497	7 537	7 577	7 617	7 658	7 698	7 739
41	7 780	7 821	7 863	7 905	7 946	7 988	8 031	8 073	8 116	8 159
42	8 202	8 245	8 288	8 332	8 375	8 419	8 464	8 508	8 553	8 597
43	8 642	8 687	8 733	8 778	8 824	8 870	8 916	8 963	9 010	9 056
44	9 103	9 151	9 198	9 246	9 294	9 342	9 390	9 439	9 487	9 536
45	9 590	9 630	9 680	9 730	9 780	9 830	9 890	9 940	9 990	10 040
46	10 090	10 140	10 190	10 240	10 300	10 350	10 400	10 460	10 510	10 560
47	10 620	10 670	10 720	10 780	10 830	10 890	10 940	11 000	11 050	11 110
48	11 170	11 220	11 280	11 340	11 390	11 450	11 510	11 570	11 620	11 680
49	11 740	11 800	11 860	11 920	11 980	12 040	12 100	12 160	12 220	12 280
50	12 340	12 400	12 460	12 520	12 590	12 650	12 710	12 770	12 840	12 900
51	12 960	13 030	13 090	13 160	13 220	13 290	13 350	13 420	13 480	13 550
52	13 620	13 680	13 750	13 820	13 890	13 950	14 020	14 090	14 160	14 230
53	14 300	14 370	14 440	14 510	14 580	14 650	14 720	14 790	14 860	14 930
54	15 010	15 080	15 150	15 230	15 300	15 370	15 450	15 520	15 600	15 670
55	15 750	15 820	15 900	15 970	16 050	16 130	16 200	16 280	16 360	16 440
56	16 520	16 590	16 670	16 750	16 830	16 910	16 990	17 070	17 150	17 240
57	17 320	17 400	17 480	17 560	17 650	17 730	17 810	17 000	17 980	18 070
58	18 150	18 240	18 320	18 410	18 500	18 580	18 670	18 760	18 850	18 930
59	19 020	19 110	19 200	19 290	19 380	19 470	19 560	19 650	19 740	19 830

Table 6.1. (cont.)

°C	0.0	0.1	0.2	0.3	0.4	0.5	0.6	0.7	0.8	0.9
60	19 930	20 020	20 110	20 200	20 300	20 390	20 490	20 580	20 680	20 770
61	20 870	20 960	21 060	21 160	21 250	21 350	21 450	21 550	21 650	21 750
62	21 840	21 940	22 040	22 150	22 250	22 350	22 450	22 550	22 650	22 760
63	22 860	22 970	23 070	23 170	23 280	23 380	23 490	23 600	23 700	23 810
64	23 920	24 030	24 130	24 240	24 350	24 460	24 570	24 680	24 790	24 900
65	25 020	25 130	25 240	25 350	25 470	25 580	25 690	25 810	25 920	26 040
66	26 160	26 270	26 390	26 510	26 620	26 740	26 860	26 980	27 100	27 220
67	27 340	27 460	27 580	27 700	27 830	27 950	28 070	28 200	28 320	28 440
68	28 570	28 700	28 820	28 950	29 070	29 200	29 330	29 460	29 590	29 720
69	29 850	29 980	30 110	30 240	30 370	30 500	30 630	30 770	30 900	30 030
70	31 170	31 300	31 400	31 580	31 710	31 850	31 990	32 120	32 260	32 400
71	32 540	32 680	32 820	32 960	33 110	33 250	33 390	33 530	33 680	33 820
72	33 970	34 110	34 260	34 400	34 550	34 700	34 840	34 990	35 140	35 290
73	35 440	35 590	35 740	35 890	36 050	36 200	36 350	36 510	36 660	36 820
74	36 970	37 130	37 280	37 440	37 600	37 760	37 920	38 070	38 230	38 390
75	38 560	38 720	38 880	39 040	39 210	39 370	39 530	39 700	39 860	40 030
76	40 200	40 370	40 530	40 700	40 870	41 040	41 210	41 380	41 550	41 730
77	41 900	42 070	42 250	42 420	42 600	42 770	42 950	43 120	43 300	43 480
78	43 660	43 840	44 020	44 200	44 380	44 560	44 740	44 930	45 110	45 300
79	45 480	45 670	45 850	46 040	46 230	46 420	46 600	46 790	46 980	47 180

80	47 370	47 560	47 750	47 940	48 140	48 330	48 530	48 730	48 920	49 120
81	49 320	49 520	49 720	49 920	50 120	50 320	50 520	50 720	50 930	51 130
82	51 340	51 540	51 750	51 950	52 160	52 370	52 580	52 790	53 000	53 210
83	53 420	53 630	53 850	54 060	54 280	54 490	54 710	54 920	55 140	55 360
84	55 580	55 800	56 020	56 240	56 460	56 690	56 910	57 130	57 360	57 580
85	57 810	58 040	58 260	58 490	58 720	58 950	59 180	59 410	59 650	59 880
86	60 110	60 350	60 580	60 820	61 060	61 290	61 530	61 770	62 010	62 250
87	62 490	62 740	62 980	63 220	63 470	63 710	63 960	64 210	64 450	64 700
88	64 950	65 200	65 450	65 710	65 960	66 210	66 470	66 720	66 980	67 230
89	67 490	67 750	68 010	68 270	68 530	68 790	69 050	69 320	69 580	69 850
90	70 110	70 380	70 650	70 920	71 190	71 460	71 730	72 000	72 270	72 540
91	72 820	73 090	73 370	73 650	73 930	74 200	74 480	74 760	75 050	75 330
92	75 610	75 900	76 180	76 470	76 750	77 040	77 330	77 620	77 910	78 200
93	78 490	78 780	79 080	79 370	79 670	79 970	80 260	80 560	80 860	81 160
94	81 460	81 770	82 070	82 370	82 680	82 980	83 290	83 600	83 910	84 220
95	84 530	84 840	85 150	85 470	85 780	86 100	86 410	86 730	87 050	87 370
96	87 690	88 010	88 330	88 650	88 980	89 300	89 630	89 960	90 290	90 610
97	90 940	91 280	91 610	91 940	92 280	92 610	92 950	93 280	93 620	93 960
98	94 300	94 640	94 990	95 330	95 670	96 020	96 360	96 710	97 060	97 410
99	97 800	98 100	98 500	98 800	99 200	99 500	99 900	100 200	100 600	101 000
100	101 300	101 700	102 100	102 400	102 800	103 100	103 500	103 900	104 300	104 600
101	105 000	105 400	105 700	106 100	106 500	106 900	107 200	107 600	108 000	108 400

6.2   Methods of Measurement

The moisture content of the atmosphere is one of the most difficult environmental variables to measure. Wexler (1970) summarized instruments known as hygrometers which are used to measure water vapor content in the free atmosphere near the earth. Those instruments that can be operated continuously or automatically under field conditions and their technical specifications are listed in Table 6.2. The following sections describe the principle of operation and construction of these instruments.

6.2.1   Psychrometer

The psychrometer is one of the oldest instruments used for determining water vapor content of the atmosphere. There are two temperature sensors in a psychrometer. The dry bulb measures air temperature and the wet bulb, covered with an evaporating water surface, measures the wet-bulb temperature. Virtually any type of temperature sensors may be used.

The psychrometric formula relates atmospheric moisture to the dry- and wet-bulb measurements. The heat given up by the air to the wet bulb in unit time is equivalent to the heat lost from a mass of air m_1 cooled from air temperature, T_a, to T_w. This heat will evaporate enough water to saturate a mass of air m_2 at the temperature T_w. This moisture gain represents a gain in latent heat. The equality between heat loss and heat gain is

$$m_1 c_p (T_a - T_w) = (e_w - e)\varepsilon m_2 \frac{L}{P}, \tag{6.14}$$

where c_p is the specific heat of air at constant pressure, ε is the ratio (0.622) of the molecular weight of water vapor to that of dry air, L is the latent heat of vaporization at the wet-bulb temperature, and e_w is the saturation vapor pressure at the wet-bulb temperature. Solving for vapor pressure gives

$$e = e_w - \frac{m_1 c_p}{m_2 \varepsilon L} P(T_a - T_w). \tag{6.15}$$

The ratio m_1/m_2 approaches unity when ventilation velocities exceed about 3 m s^{-1}. The thermodynamic psychrometric constant $A = c_p(\varepsilon L)^{-1}$ equals $0.000\,646°C^{-1}$ at $0°C$ and varies with temperature as the latent heat of vaporization (J g^{-1}) varies with temperature,

$$L = 2\,500.25 - 2.365\,T_w, \tag{6.16}$$

or

$$L = 2\,500.25(1 - 0.000\,946\,T_w). \tag{6.17}$$

An alternate expression for the thermodynamic psychrometric constant is

$$A = 0.000\,646(1 + 0.000\,946\,T_w). \tag{6.18}$$

Table 6.2. Summary of characteristics of humidity sensors that can be operated continuously or automatically under field conditions (after Wexler, 1970).

Type of hygrometer	Range			Uncertainty		Ambient range (°C)	Response time	Primary output parameter
	Mixing ratio (g/kg)	Dew point (°C)	Relative humidity (%)	Dew point (°C)	Relative humidity (%)			
Aspirated psychrometer			5 to 100		1	5 to 65	medium to fast	temperature
Hair (human blond)			2 to 100		3	−50 to 60	medium	elongation
Dew point	0.003 to 622	−80 to 100		0.1			medium to fast	temperature
Dunmore			7 to 98		1.5	5 to 60	fast	resistance
Jason			25 to 85		5	−40 to +40	fast	resistance
Carbon			10 to 100		5	−40 to +40	fast	resistance
Ion exchange			10 to 100		5	−40 to +95	fast	resistance
Dewcel	0.23 to 622	−30 to 100		1.5		−30 to 100	medium	temperature
Ionic crystal	0.016 to 15	−55 to 20		0.05		room	fast	temperature
Infrared	0.045 to 15	−45 to 20		3 to 0.5		field		IR radiation
Microwave	0.016 to 49	−55 to 40		0.7 to 0.1		field	fast	frequency
Thermal conductivity	0.76 to 87	−18 to 50		1 to 0.1		field		temperature

Table 6.3. Slope of the saturation vapor pressure curve, Δ; latent heat of vaporization, L; psychrometric constant at atmospheric pressure ($P = 100\,\text{kPa}$), γ; Δ/γ; and $\Delta/(\Delta + \gamma)$ for various temperatures T.

T	Δ	L	γ	$\dfrac{\Delta}{\gamma}$	$\dfrac{\Delta}{\Delta + \gamma}$
(°C)	(Pa °C^{-1})	(J g^{-1})	(Pa °C^{-1})		
−9	24.3	2 522	64.1	0.379	0.275
−8	26.0	2 520	64.1	0.406	0.289
−7	27.9	2 517	64.2	0.435	0.303
−6	29.9	2 515	64.2	0.465	0.317
−5	32.0	2 513	64.3	0.497	0.332
−4	34.2	2 510	64.4	0.531	0.347
−3	36.5	2 508	64.4	0.567	0.362
−2	39.0	2 506	64.5	0.605	0.377
−1	41.6	2 503	64.5	0.645	0.392
0	44.4	2 501	64.6	0.688	0.408
1	47.4	2 499	64.7	0.733	0.423
2	50.5	2 496	64.7	0.780	0.438
3	53.7	2 494	64.8	0.830	0.453
4	57.2	2 491	64.8	0.882	0.469
5	60.9	2 489	64.9	0.938	0.484
6	64.7	2 487	65.0	0.996	0.499
7	68.8	2 484	65.0	1.057	0.514
8	73.0	2 482	65.1	1.122	0.529
9	77.5	2 480	65.2	1.190	0.543
10	82.2	2 477	65.2	1.261	0.558
11	87.2	2 475	65.3	1.336	0.572
12	92.4	2 473	65.3	1.415	0.586
13	97.9	2 470	65.4	1.497	0.600
14	103.7	2 468	65.5	1.584	0.613
15	109.7	2 465	65.5	1.675	0.626
16	116.1	2 463	65.6	1.770	0.639
17	122.7	2 461	65.7	1.869	0.651
18	129.7	2 458	65.7	1.974	0.664
19	137.0	2 456	65.8	2.083	0.676
20	144.7	2 454	65.8	2.197	0.687
21	152.7	2 451	65.9	2,316	0.698
22	161.0	2 449	66.0	2.441	0.709
23	169.8	2 447	66.0	2.572	0.720
24	179.0	2 444	66.1	2.708	0.730
25	188.6	2 442	66.2	2.850	0.740

Table 6.3. (*cont.*)

T	Δ	L	γ	$\dfrac{\Delta}{\gamma}$	$\dfrac{\Delta}{\Delta + \gamma}$
(°C)	(Pa °C^{-1})	(J g^{-1})	(Pa °C^{-1})		
26	198.6	2 439	66.2	2.999	0.750
27	209.0	2 437	66.3	3.153	0.759
28	219.9	2 435	66.4	3.315	0.768
29	231.3	2 432	66.4	3.483	0.777
30	243.2	2 430	66.5	3.658	0.785
31	255.6	2 428	66.5	3.841	0.793
32	268.5	2 425	66.6	4.031	0.801
33	282.0	2 423	66.7	4.229	0.809
34	296.0	2 420	66.7	4.434	0.816
35	310.6	2 418	66.8	4.648	0.823
36	325.7	2 416	66.9	4.871	0.830
37	341.5	2 413	66.9	5.102	0.836
38	358.0	2 411	67.0	5.342	0.842
39	375.1	2 409	67.1	5.592	0.848

The psychrometric constant, in more common use than the thermodynamic psychrometric constant, includes atmospheric pressure in its definition, $\gamma = AP$. Additional values of γ are given in Table 6.3.

Ferrel (1886), according to Harrison (1963), determined an empirical psychrometric constant from whirling a psychrometer (wet bulb 23 mm long and 4 mm diameter) at 7.6 m s^{-1}. His "so-called psychrometric constant" is

$$A_o = 0.000\,660°C^{-1}. \tag{6.19}$$

Taking into account the temperature coefficient of L, his psychrometric equation is

$$e = e_w - 0.000\,660(1 + 0.001\,15\,T_w)(T_a - T_w)P. \tag{6.20}$$

This form has been used by the U.S. Weather Bureau (Marvin, 1941).

The vapor for an atmospheric pressure of 100 and 85 kPa as a function of the wet-bulb temperature, T_w, and the depression between air and wet-bulb temperatures, DT_w, is given in Tables 6.4 and 6.5. The percent relative humidity for 100 and 85 kPa as a function of wet-bulb temperature and wet-bulb depression is given in Tables 6.6 and 6.7.

Errors in results from psychrometers can arise from several sources: unmatched thermometers, atmospheric pressure, radiation, conduction, inadequate ventilation and impure water. Errors in wet-bulb depression,

Table 6.4. Vapor pressure (Pa) at 100 kPa pressure for various wet-bulb temperatures (°C) and wet-bulb depressions, DT_w (°C). When temperature is below 0°C, saturation vapor pressure is computed for ice.

DT_w	Wet-bulb temperature									
	−10	−9	−8	−7	−6	−5	−4	−3	−2	−1
0	260	284	310	338	368	402	437	476	517	562
1	190	240	270	300	330	360	390	420	460	500
2	130	180	200	230	260	290	320	360	400	440
3	060	110	140	170	190	230	260	290	330	370
4		050	070	100	130	160	190	230	270	310
5			010	030	060	090	130	160	200	240
6						030	060	100	130	170
7								030	070	110
										040

	0	1	2	3	4	5	6	7	8	9
0	611	657	705	758	813	872	935	1 001	1 072	1 147
1	545	591	639	691	747	806	869	935	1 006	1 081
2	478	524	573	625	681	740	802	869	940	1 015
3	412	458	507	559	614	673	736	803	873	949
4	346	391	440	492	548	607	669	736	807	882
5	279	325	374	426	481	540	603	669	740	815
6	212	258	307	359	414	473	536	603	673	749
7	145	191	240	292	347	406	469	536	606	682
8	078	124	173	225	280	339	402	468	539	615
9	011	056	105	157	213	272	334	401	472	547
10			038	090	145	204	267	334	405	480
11				022	078	137	199	266	337	412
12					010	069	132	198	269	344
13						001	064	130	201	277
14								062	133	209
15									065	140
16										072
17										003

Table 6.4. (*cont.*)

DT_w	\multicolumn{10}{c}{Wet-bulb temperature}
	10	11	12	13	14	15	16	17	18	19
0	1 227	1 312	1 402	1 497	1 598	1 704	1 817	1 937	2 063	2 196
1	1 160	1 245	1 335	1 430	1 531	1 638	1 750	1 870	1 996	2 130
2	1 093	1 178	1 268	1 363	1 464	1 571	1 683	1 803	1 929	2 063
3	1 026	1 111	1 201	1 296	1 397	1 503	1 616	1 736	1 862	1 995
4	959	1 044	1 133	1 229	1 329	1 436	1 549	1 668	1 795	1 928
5	892	976	1 066	1 161	1 262	1 369	1 482	1 601	1 727	1 861
6	824	909	998	1 094	1 194	1 301	1 414	1 533	1 660	1 793
7	756	841	931	1 026	1 127	1 233	1 346	1 466	1 592	1 725
8	688	773	863	958	1 059	1 165	1 278	1 398	1 524	1 657
9	620	705	795	890	991	1 097	1 210	1 330	1 456	1 589
10	552	637	727	822	922	1 029	1 142	1 262	1 388	1 521
11	484	568	658	753	854	961	1 074	1 193	1 319	1 453
12	415	500	590	685	786	892	1 005	1 125	1 251	1 384
13	347	431	521	616	717	824	937	1 056	1 182	1 316
14	278	362	452	547	648	755	868	987	1 113	1 247
15	209	293	383	478	579	686	799	918	1 045	1 178
16	140	224	314	409	510	617	730	849	975	1 109
17	070	155	245	340	441	548	660	780	906	1 040
18	001	086	175	271	371	478	591	710	837	970
19		016	106	201	302	409	521	641	767	901
20			036	131	232	339	452	571	697	831
21			061	162	269	382	501	628	761
22				092	199	312	431	558	691
23				022	129	242	361	487	621
24					058	171	291	417	550
25						101	220	347	480
26						030	150	276	409
27							079	205	339
28							008	134	268
29								063	197
30									125
31									054

Table 6.4. (*cont.*)

DT_w	Wet-bulb temperature									
	20	21	22	23	24	25	26	27	28	29
0	2 337	2 486	2 643	2 809	2 983	3 167	3 361	3 565	3 780	4 005
1	2 270	2 418	2 575	2 741	2 916	3 099	3 293	3 497	3 712	3 938
2	2 220	2 351	2 508	2 673	2 848	3 032	3 226	3 430	3 644	3 870
3	2 134	2 283	2 440	2 605	2 780	2 964	3 158	3 362	3 576	3 802
4	2 066	2 215	2 372	2 537	2 712	2 896	3 090	3 294	3 508	3 734
5	1 998	2 147	2 304	2 469	2 644	2 828	3 021	3 225	3 440	3 666
6	1 929	2 078	2 235	2 401	2 575	2 759	2 953	3 157	3 372	3 598
7	1 861	2 010	2 167	2 332	2 507	2 691	2 885	3 089	3 303	3 529
8	1 792	1 941	2 098	2 264	2 438	2 622	2 816	3 020	3 235	3 460
9	1 723	1 872	2 029	2 195	2 369	2 553	2 747	2 951	3 166	3 392
10	1 655	1 803	1 960	2 126	2 300	2 484	2 678	2 882	3 097	3 323
11	1 585	1 734	1 891	2 057	2 231	2 415	2 609	2 813	3 028	3 254
12	1 516	1 665	1 822	1 987	2 162	2 346	2 540	2 744	2 958	3 184
13	1 447	1 595	1 752	1 918	2 093	2 277	2 470	2 674	2 889	3 115
14	1 377	1 526	1 683	1 848	2 023	2 207	2 401	2 605	2 819	3 045
15	1 307	1 456	1 613	1 770	1 953	2 137	2 331	2 535	2 750	2 976
16	1 238	1 386	1 543	1 709	1 883	2 067	2 261	2 465	2 680	2 906
17	1 168	1 316	1 473	1 639	1 813	1 997	2 191	2 395	2 610	2 836
18	1 097	1 246	1 403	1 569	1 743	1 927	2 121	2 325	2 540	2 766
19	1 027	1 176	1 333	1 498	1 673	1 857	2 051	2 255	2 469	2 695
20	957	1 105	1 262	1 428	1 602	1 786	1 980	2 184	2 399	2 625
21	886	1 035	1 192	1 357	1 532	1 716	1 909	2 114	2 328	2 554
22	815	964	1 121	1 286	1 461	1 645	1 839	2 043	2 257	2 483
23	744	893	1 050	1 216	1 390	1 574	1 768	1 972	2 187	2 412
24	673	822	979	1 144	1 319	1 503	1 697	1 901	2 115	2 341
25	602	751	908	1 073	1 248	1 432	1 625	1 829	2 044	2 270
26	531	679	836	1 002	1 176	1 360	1 554	1 758	1 973	2 199
27	459	608	765	930	1 105	1 289	1 483	1 687	1 901	2 127
28	387	536	693	859	1 033	1 217	1 411	1 615	1 830	2 055
29	315	464	621	787	961	1 145	1 339	1 543	1 758	1 984
30	243	392	549	715	889	1 073	1 267	1 471	1 686	1 912
31	171	320	477	643	817	1 001	1 195	1 399	1 614	1 839
32	99	248	405	570	745	929	1 123	1 327	1 541	1 767
33	27	175	332	498	672	856	1 050	1 254	1 469	1 695
34		103	260	425	600	784	977	1 182	1 396	1 622
35		30	187	352	527	711	905	1 109	1 323	1 549
36			114	280	454	638	832	1 036	1 251	1 476
37			41	207	381	565	759	963	1 178	1 403
38				133	308	492	686	890	1 104	1 330
39				60	234	418	612	816	1 031	1 257
40					161	345	539	743	957	1 183

Table 6.4. (*cont.*)

DT_w	Wet-bulb temperature									
	30	31	32	33	34	35	36	37	38	39
0	4 243	4 493	4 755	5 031	5 320	5 624	5 942	6 276	6 626	6 993
1	4 175	4 424	4 687	4 962	5 252	5 555	5 874	6 208	6 558	6 925
2	4 106	4 356	4 618	4 894	5 183	5 487	5 805	6 139	6 490	6 857
3	4 038	4 287	4 550	4 825	5 114	5 418	5 737	6 071	6 421	6 788
4	3 669	4 218	4 481	4 756	5 046	5 349	5 668	6 002	6 352	6 719
5	3 900	4 149	4 412	4 687	4 977	5 280	5 599	5 933	6 283	6 650
6	3 831	4 080	4 343	4 618	4 908	5 211	5 530	5 864	6 214	6 581
7	3 761	4 011	4 273	4 549	4 838	5 142	5 461	5 795	6 145	6 512
8	3 692	3 942	4 204	4 480	4 769	5 073	5 391	5 725	6 075	6 442
9	3 622	3 872	4 134	4 410	4 699	5 003	5 322	5 656	6 006	6 373
10	3 553	3 802	4 065	4 340	4 630	4 933	5 252	5 586	5 936	6 303
11	3 483	3 733	3 995	4 270	4 560	5 863	5 182	5 516	5 866	6 233
12	3 413	3 662	3 925	4 200	4 490	4 793	5 112	5 446	5 796	6 163
13	3 343	3 592	3 855	4 130	4 420	4 723	5 042	5 376	5 726	6 093
14	3 272	3 522	3 784	4 060	4 349	4 653	4 971	5 305	5 656	6 023
15	3 202	3 452	3 714	3 989	4 279	4 582	4 901	5 235	5 585	5 952
16	3 131	3 381	3 643	3 919	4 208	4 512	4 830	5 164	5 515	5 882
17	3 060	3 310	3 572	3 848	4 137	4 441	4 760	6 094	5 444	5 811
18	2 939	3 239	3 502	3 777	4 066	4 370	4 689	5 023	5 373	5 740
19	2 918	3 168	3 430	3 706	3 995	4 299	4 618	4 952	5 302	5 669
20	2 847	3 097	3 359	3 635	3 924	4 228	4 546	4 880	5 231	5 597
21	2 776	3 025	3 288	3 563	3 853	4 156	4 476	4 809	5 159	5 526
22	2 704	2 954	3 216	3 492	3 781	4 085	4 403	4 737	5 088	5 455
23	2 633	2 882	3 145	3 420	3 709	4 013	4 332	4 666	5 016	5 383
24	2 561	2 810	3 073	3 348	3 638	3 941	4 260	4 594	4 944	5 311
25	2 489	2 738	3 001	3 276	3 566	3 869	4 188	4 522	4 872	5 239
26	2 417	2 666	2 929	3 204	3 494	3 797	4 116	4 450	4 800	5 167
27	2 344	2 594	2 856	3 132	3 421	3 725	4 043	4 377	4 728	5 095
28	2 272	2 521	2 784	3 059	3 349	3 652	3 971	4 305	4 655	5 022
29	2 199	2 449	2 711	2 987	3 276	3 580	3 898	4 232	4 583	4 950
30	2 126	2 376	2 638	2 914	3 203	3 507	3 826	4 160	4 510	4 877
31	2 054	2 303	2 566	2 841	3 130	3 434	3 753	4 087	4 437	4 804
32	1 980	2 230	2 493	2 768	3 057	3 361	3 680	4 014	4 364	4 731
33	1 907	2 157	2 419	2 695	2 984	3 288	3 606	3 940	4 291	4 658
34	1 834	2 084	2 346	2 622	2 911	3 214	3 533	3 867	4 217	4 584
35	1 760	2 010	2 272	2 548	2 837	3 141	3 460	3 794	4 144	4 511
36	1 687	1 936	2 199	2 474	2 764	3 067	3 386	3 720	4 070	4 437
37	1 613	1 863	2 125	2 401	2 690	2 993	3 312	3 646	3 996	4 363
38	1 539	1 789	2 051	2 327	2 616	2 920	3 238	3 572	3 922	4 289
39	1 465	1 714	1 977	3 363	2 542	2 845	3 164	3 498	3 848	4 215
40	1 391	1 640	1 903	2 178	2 467	2 771	3 090	3 424	3 774	4 141

Table 6.5. Vapor pressure (Pa) at 85 kPa pressure for various wet-bulb temperatures (°C) and wet-bulb depressions, DT_w (°C). When temperature is below 0°C, saturation vapor pressure is computed for ice.

DT_w	Wet-bulb temperature									
	-10	-9	-8	-7	-6	-5	-4	-3	-2	-1
0	260	284	310	338	368	402	437	476	517	562
1	200	250	280	310	340	370	400	430	470	510
2	150	200	220	250	280	310	340	380	420	460
3	090	140	170	190	220	250	290	320	360	400
4	040	090	110	140	170	200	230	270	300	340
5		030	060	080	110	140	180	210	250	290
6			030	060	090	120	150	190	230
7					030	060	100	140	180
8						010	040	80	120
9								020	60
10									010

	0	1	2	3	4	5	6	7	8	9
0	611	657	705	758	813	872	935	1 001	1 072	1 147
1	555	600	649	701	757	816	878	945	1 016	1 091
2	498	544	593	645	700	759	822	989	960	1 035
3	443	488	537	589	644	703	766	832	903	979
4	385	431	480	532	588	646	709	776	847	922
5	329	375	423	475	531	590	653	719	790	865
6	272	318	367	419	474	533	596	662	733	808
7	215	261	310	362	417	476	539	606	676	752
8	158	204	253	305	360	419	482	548	619	694
9	101	146	195	247	303	362	425	491	562	637
10	043	089	138	190	245	304	367	434	505	580
11		032	081	133	188	247	310	376	447	522
12			023	075	130	189	252	319	390	465
13				017	073	132	194	261	332	407
14					015	074	137	203	274	349
15						016	079	145	216	291
16							021	087	158	233
17								029	100	175
18									041	117
19										058

Table 6.5. (*cont.*)

DT_w	Wet-bulb temperature									
	10	11	12	13	14	15	16	17	18	19
0	1 227	1 312	1 402	1 497	1 598	1 704	1 817	1 937	2 063	2 196
1	1 170	1 255	1 345	1 440	1 541	1 648	1 760	1 880	2 006	2 140
2	1 113	1 198	1 288	1 383	1 484	1 591	1 704	1 823	1 949	2 083
3	1 056	1 141	1 231	1 326	1 427	1 534	1 646	1 766	1 892	2 026
4	999	1 084	1 174	1 269	1 370	1 476	1 589	1 709	1 835	1 968
5	942	1 027	1 116	1 212	1 312	1 419	1 532	1 651	1 778	1 911
6	884	969	1 059	1 154	1 255	1 362	1 475	1 594	1 720	1 854
7	827	912	1 001	1 096	1 197	1 304	1 417	1 536	1 663	1 796
8	769	854	944	1 039	1 140	1 246	1 359	1 479	1 605	1 738
9	711	796	886	981	1 082	1 188	1 301	1 421	1 547	1 680
10	653	738	828	923	1 024	1 130	1 243	1 363	1 489	1 622
11	595	680	770	865	966	1 072	1 185	1 305	1 431	1 564
12	537	622	711	807	907	1 014	1 127	1 246	1 373	1 506
13	470	563	653	748	849	956	1 069	1 188	1 314	1 448
14	420	505	595	690	791	897	1 010	1 130	1 256	1 389
15	362	446	536	631	732	839	952	1 071	1 197	1 331
16	303	387	477	572	673	780	893	1 012	1 139	1 272
17	244	329	418	514	614	721	834	953	1 080	1 213
18	185	270	359	455	555	662	775	894	1 021	1 154
19	125	210	300	395	496	603	716	835	962	1 095
20	067	151	241	336	437	544	657	776	902	1 036
21	007	092	182	277	378	484	597	717	843	976
22		032	122	217	318	425	538	657	783	917
23			062	158	258	365	478	597	724	857
24			003	098	199	305	418	538	664	797
25				038	139	245	358	478	604	737
26					079	185	298	418	544	677
27					019	125	238	358	484	617
28						065	178	297	424	557
29						005	117	237	363	497
30							057	176	303	436
31								116	242	375
32								055	181	314
33									120	254
34									059	191
35										131
36										070
37										008

Table 6.5. (*cont.*)

DT_w	Wet-bulb temperature									
	20	21	22	23	24	25	26	27	28	29
0	2 337	2 486	2 643	2 809	2 983	3 167	3 361	3 565	3 780	4 005
1	2 280	2 429	2 586	2 751	2 926	3 110	3 303	3 507	3 722	3 948
2	2 222	2 371	2 528	2 694	2 868	3 052	3 246	3 450	3 665	3 890
3	2 165	2 313	2 470	2 636	2 810	2 994	3 188	3 392	3 607	3 833
4	2 107	2 255	2 412	2 578	2 753	2 936	3 130	3 334	3 549	3 775
5	2 049	2 197	2 354	2 520	2 695	2 879	3 072	3 276	3 491	3 717
6	1 991	2 139	2 296	2 462	2 636	2 820	3 014	3 218	3 433	3 659
7	1 932	2 081	2 238	2 404	2 578	2 762	2 956	3 160	3 375	3 601
8	1 874	2 023	2 180	2 345	2 520	2 704	2 898	3 102	3 316	3 542
9	1 816	1 964	2 121	2 287	2 461	2 645	2 839	3 043	3 258	3 484
10	1 757	1 906	2 063	2 228	2 403	2 587	2 780	2 985	3 199	3 425
11	1 698	1 847	2 004	2 169	2 344	2 528	2 722	2 926	3 140	3 366
12	1 639	1 788	1 945	2 111	2 285	2 469	2 663	2 867	3 082	3 308
13	1 580	1 729	1 886	2 052	2 226	2 410	2 604	2 808	3 023	3 249
14	1 521	1 670	1 827	1 992	2 167	2 351	2 545	2 749	2 963	3 189
15	1 462	1 611	1 768	1 933	2 108	2 292	2 485	2 689	2 904	3 130
16	1 403	1 551	1 708	1 874	2 048	2 232	2 426	2 630	2 845	3 071
17	1 343	1 492	1 649	1 814	1 989	2 173	2 367	2 571	2 785	3 011
18	1 283	1 432	1 589	1 755	1 929	2 113	2 307	2 511	2 726	2 952
19	1 224	1 372	1 529	1 695	1 869	2 053	2 247	2 451	2 666	2 892
20	1 164	1 312	1 469	1 635	1 810	1 993	2 187	2 391	2 606	2 832
21	1 104	1 252	1 409	1 575	1 749	1 933	2 127	2 331	2 546	2 772
22	1 043	1 192	1 349	1 515	1 689	1 873	2 067	2 271	2 486	2 712
23	983	1 132	1 289	1 454	1 629	1 813	2 007	2 211	2 425	2 651
24	923	1 072	1 228	1 394	1 569	1 753	1 946	2 150	2 365	2 591
25	862	1 011	1 168	1 334	1 508	1 692	1 886	2 090	2 305	2 530
26	802	950	1 107	1 273	1 447	1 631	1 825	2 029	2 244	2 470
27	741	889	1 046	1 212	1 387	1 571	1 764	1 968	2 183	2 409
28	680	829	986	1 151	1 326	1 510	1 703	1 907	2 122	2 348
29	619	767	924	1 090	1 265	1 449	1 642	1 846	2 061	2 287
30	558	706	863	1 029	1 203	1 387	1 581	1 785	2 000	2 226
31	496	645	802	967	1 142	1 326	1 520	1 724	1 938	2 164
32	435	583	740	906	1 081	1 265	1 458	1 662	1 877	2 103
33	373	522	679	844	1 019	1 203	1 397	1 601	1 815	2 041
34	311	460	617	783	957	1 141	1 335	1 539	1 754	1 980
35	250	398	555	721	895	1 079	1 273	1 477	1 692	1 918
36	188	336	493	659	833	1 017	1 211	1 415	1 630	1 856
37	126	274	431	597	771	955	1 149	1 353	1 568	1 794
38	063	212	369	535	709	893	1 087	1 291	1 506	1 731
39	001	150	307	472	647	831	1 024	1 229	1 443	1 669
40		087	244	410	584	768	962	1 166	1 381	1 607

Table 6.5. (*cont.*)

DT_w	Wet-bulb temperature									
	30	31	32	33	34	35	36	37	38	39
0	4 243	4 493	4 755	5 031	5 320	5 624	5 942	6 276	6 626	6 993
1	4 185	4 435	4 697	4 973	5 262	5 566	5 884	6 218	6 568	6 935
2	4 127	4 376	4 639	4 914	5 204	5 507	5 826	6 160	6 510	6 877
3	4 068	4 318	4 580	4 856	5 145	5 449	5 768	6 102	6 452	6 819
4	4 010	4 260	4 522	4 798	5 087	5 390	5 709	6 043	6 393	6 760
5	3 951	4 201	4 463	4 739	5 028	5 332	5 650	5 984	6 335	6 702
6	3 892	4 142	4 405	4 680	4 969	5 273	5 592	5 926	6 276	6 643
7	3 834	4 083	4 346	4 621	4 911	5 214	5 533	5 867	6 217	6 584
8	3 775	4 024	4 287	4 562	4 852	5 155	5 474	5 808	6 158	6 525
9	3 715	3 965	4 228	4 503	4 792	5 096	5 415	5 749	6 099	6 466
10	3 656	3 906	4 168	4 444	4 733	5 037	5 355	5 689	6 040	6 407
11	3 597	3 847	4 109	4 384	4 674	4 977	5 296	5 630	5 980	6 347
12	3 537	3 787	4 049	4 325	4 614	4 918	5 236	5 571	5 921	6 288
13	3 478	3 727	3 990	4 265	4 555	4 858	5 177	5 511	5 861	6 228
14	3 417	3 668	3 930	4 206	4 495	4 799	5 117	5 451	5 801	6 168
15	3 358	3 608	3 870	4 146	4 435	4 739	5 057	5 391	5 741	6 108
16	3 298	3 548	3 810	4 086	4 375	4 679	4 997	5 331	5 681	6 048
17	3 238	3 487	3 750	4 025	4 315	4 618	4 937	5 271	5 621	5 988
18	3 177	3 427	3 690	3 965	4 264	4 558	4 877	5 211	5 561	5 928
19	3 117	3 367	3 629	3 905	4 194	4 498	4 816	5 150	5 500	5 867
20	3 057	3 306	3 569	3 844	4 133	4 437	4 756	5 090	5 440	5 807
21	2 996	3 246	3 508	3 783	4 073	4 376	4 695	5 029	5 379	5 746
22	2 935	3 185	3 447	3 723	4 012	4 316	4 634	4 968	5 318	5 685
23	2 874	3 124	3 386	3 662	3 951	4 255	4 573	4 907	5 257	5 624
24	2 813	3 063	3 325	3 601	3 890	4 194	4 512	4 846	5 196	5 563
25	2 752	3 002	3 264	3 539	3 829	4 132	4 451	4 785	5 135	5 502
26	2 690	2 940	3 203	3 478	3 767	4 071	4 390	4 724	5 074	5 441
27	2 629	2 879	3 141	3 417	3 706	4 010	4 328	4 662	5 012	5 379
28	2 567	2 816	3 080	3 355	3 644	3 948	4 267	4 601	4 951	5 318
29	2 506	2 755	3 018	3 293	3 583	3 886	4 205	4 539	4 889	5 256
30	2 444	2 594	2 956	3 232	3 521	3 825	4 143	4 477	4 827	5 194
31	2 382	2 632	2 894	3 170	3 459	3 763	4 081	4 415	4 765	5 132
32	2 320	2 570	2 832	3 107	3 397	3 700	4 019	4 353	4 703	5 070
33	2 258	2 507	2 770	3 045	3 335	3 638	3 957	4 291	4 641	5 008
34	2 195	2 445	2 707	2 983	3 272	3 576	3 894	4 228	4 579	4 946
35	2 133	2 382	2 645	2 920	3 210	3 513	3 832	4 166	4 516	4 883
36	2 070	2 320	2 582	2 858	3 147	3 451	3 769	4 103	4 454	4 820
37	2 007	2 257	2 519	2 795	3 084	3 388	3 707	4 041	4 391	4 758
38	1 945	2 194	2 457	2 732	3 022	3 325	3 644	3 978	4 328	4 695
39	1 882	2 131	2 394	2 669	2 959	3 262	3 581	3 915	4 265	4 632
40	1 818	2 068	2 330	2 606	3 895	3 199	3 518	3 852	4 202	4 569

Table 6.6. Percent relative humidity at 100 kPa pressure for various wet-bulb temperatures (°C) and wet bulb depressions, DT_w (°C). When temperature is below 0°C, saturation vapor pressure is computed for ice.

DT_w	Wet-bulb temperature									
	-10	-9	-8	-7	-6	-5	-4	-3	-2	-1
0	100.0	100.0	100.0	100.0	100.0	100.0	100.0	100.0	100.0	100.0
1	62.8	73.0	74.5	75.9	77.2	78.4	79.5	80.5	81.4	82.3
2	38.5	49.5	52.2	54.8	57.2	59.3	61.4	63.2	65.0	66.6
3	17.5	29.0	32.8	36.4	39.6	42.7	45.5	48.0	50.4	52.6
4		11.3	16.0	20.3	24.3	28.1	31.5	34.7	37.6	40.3
5			1.4	6.4	11.0	15.3	19.3	22.9	26.3	29.5
6						4.2	8.6	12.6	16.4	19.9
7								3.6	7.7	11.5
8									0.1	4.1

DT_w	0	1	2	3	4	5	6	7	8	9
0	100.0	100.0	100.0	100.0	100.0	100.0	100.0	100.0	100.0	100.0
1	83.0	83.7	84.4	85.1	85.7	86.2	86.7	87.2	87.7	88.1
2	67.8	69.2	70.5	71.7	72.8	73.9	74.8	75.7	76.6	77.4
3	54.4	56.3	58.1	59.8	61.3	62.8	64.1	65.4	66.6	67.7
4	42.5	44.9	47.1	49.2	51.1	52.9	54.5	56.1	57.6	58.9
5	32.0	34.7	37.3	39.7	41.9	44.0	45.9	47.8	49.5	51.0
6	22.7	25.8	28.6	31.3	33.8	36.1	38.2	40.3	42.2	43.9
7	14.5	17.8	20.9	23.8	26.5	29.0	31.3	33.5	35.6	37.5
8	7.3	10.8	14.1	17.1	20.0	22.7	25.1	27.5	29.7	31.7
9	0.9	4.6	8.0	11.2	14.2	17.0	19.6	22.1	24.4	26.5
10			2.7	6.0	9.1	12.0	14.7	17.2	19.6	21.8
11				1.4	4.6	7.5	10.3	12.9	15.3	17.6
12					0.6	3.6	6.4	9.0	11.5	13.9
13						0.1	2.9	5.6	8.1	10.5
14								2.5	5.0	7.4
15									2.3	4.7
16										2.3
17										0.1

Table 6.6. (*cont.*)

DT_w	Wet-bulb temperature									
	10	11	12	13	14	15	16	17	18	19
0	100.0	100.0	100.0	100.0	100.0	100.0	100.0	100.0	100.0	100.0
1	88.4	88.8	89.2	89.5	89.8	90.1	90.4	90.6	90.9	91.1
2	78.0	78.7	79.4	80.0	80.6	81.1	81.6	82.1	82.5	83.0
3	68.6	69.5	70.5	71.3	72.1	72.9	73.6	74.3	74.9	75.5
4	60.0	61.2	62.4	63.4	64.4	65.4	66.3	67.1	67.9	68.7
5	52.3	53.7	55.0	56.3	57.5	58.6	59.6	60.6	61.5	62.4
6	45.3	46.9	48.4	49.0	51.1	52.3	53.5	54.6	55.6	56.6
7	39.0	40.8	42.4	43.9	45.3	46.7	47.9	49.1	50.3	51.3
8	33.4	35.2	36.9	38.5	40.1	41.5	42.9	44.1	45.3	46.5
9	28.2	30.2	32.0	33.7	35.3	36.8	38.2	39.6	40.8	42.1
10	23.6	25.6	27.5	29.3	30.9	32.5	34.0	35.4	36.7	38.0
11	19.5	21.5	23.4	25.3	27.0	28.6	30.1	31.6	32.9	34.2
12	15.7	17.8	19.8	21.6	23.4	25.0	26.6	28.1	29.5	30.8
13	12.3	14.5	16.5	18.3	20.1	21.8	23.4	24.9	26.3	27.7
14	9.3	11.4	13.5	15.4	17.1	18.8	20.5	22.0	23.4	24.8
15	6.6	8.7	10.8	12.7	14.5	16.2	17.8	19.3	20.8	22.1
16	4.2	6.3	8.3	10.2	12.0	13.7	15.3	16.9	18.3	19.7
17	2.0	4.1	6.1	8.0	9.8	11.5	13.1	14.7	16.1	17.5
18	0.0	2.1	4.1	6.0	7.8	9.5	11.1	12.6	14.1	15.5
19		0.4	2.4	4.2	6.0	7.7	9.3	10.8	12.2	13.6
20			0.8	2.6	4.4	6.0	7.6	9.1	10.5	11.9
21				1.2	2.9	4.5	6.1	7.6	9.0	10.3
22				1.6	3.2	4.7	6.2	7.6	8.9
23					0.4	1.9	3.5	4.9	6.3	7.6
24						0.8	2.3	3.7	5.1	6.4
25							1.3	2.7	4.0	5.3
26							0.4	1.7	3.0	4.3
27								0.9	2.1	3.4
28								0.1	1.3	2.5
29									0.6	1.8
30										1.1
31										0.4

Table 6.6. (*cont.*)

DT_w	Wet-bulb temperature									
	20	21	22	23	24	25	26	27	28	29
0	100.0	100.0	100.0	100.0	100.0	100.0	100.0	100.0	100.0	100.0
1	91.3	91.5	91.7	91.9	92.1	92.2	92.4	92.5	92.7	92.8
2	83.3	83.7	84.1	84.4	84.7	85.0	85.3	85.6	85.9	86.1
3	76.0	76.5	77.0	77.5	78.0	78.4	78.8	79.2	79.6	80.0
4	69.3	69.9	70.6	71.1	71.7	72.3	72.3	73.3	73.8	74.2
5	63.1	63.9	64.6	65.3	66.0	66.6	67.2	67.8	68.4	68.9
6	57.4	58.3	59.1	59.9	60.7	61.4	62.1	62.8	63.4	64.0
7	52.2	53.2	54.1	55.0	55.8	56.6	57.3	58.1	58.7	59.4
8	47.4	48.5	49.4	50.4	51.3	52.1	52.9	53.7	54.4	55.1
9	43.0	44.1	45.2	46.2	47.1	48.0	48.8	49.7	50.4	51.2
10	39.0	40.1	41.2	42.3	43.2	44.2	45.1	45.9	46.7	47.5
11	35.3	36.5	37.6	38.7	39.7	40.6	41.6	42.5	43.3	44.1
12	31.9	33.1	34.2	35.3	36.4	37.4	38.3	39.2	40.1	40.9
13	28.8	30.0	31.2	32.3	33.3	34.4	35.3	36.2	37.1	38.0
14	25.9	27.1	28.3	29.5	30.5	31.6	32.5	33.5	34.4	35.2
15	23.2	24.5	25.7	26.8	27.9	29.0	30.0	30.9	31.8	32.7
16	20.8	22.1	23.3	24.4	25.5	26.6	27.6	28.5	29.4	30.3
17	18.6	19.9	21.1	22.2	23.3	24.4	25.4	26.3	27.2	28.1
18	16.6	17.8	19.0	20.2	21.3	22.3	23.3	24.3	25.2	26.1
19	14.7	15.9	17.1	18.3	19.4	20.4	21.4	22.3	23.3	24.1
20	13.0	14.2	15.4	16.5	17.6	18.6	19.6	20.6	21.5	22.4
21	11.4	12.6	13.8	14.9	16.0	17.0	18.0	18.0	19.8	20.7
22	9.9	11.2	12.3	13.4	14.5	15.5	16.5	17.4	18.3	19.2
23	8.6	9.8	11.0	12.0	13.1	14.1	15.1	16.0	16.9	17.7
24	7.4	8.6	9.7	10.8	11.8	12.8	13.8	14.7	15.5	16.4
25	6.3	7.4	8.5	9.6	10.6	11.6	12.5	13.4	14.3	15.1
26	5.3	6.4	7.5	8.5	9.5	10.5	11.4	12.3	13.1	14.0
27	4.3	5.4	6.5	7.5	8.5	9.5	10.4	11.2	12.1	12.9
28	3.5	4.6	5.6	6.6	7.6	8.5	9.4	10.3	11.1	11.9
29	2.7	3.8	4.8	5.8	6.7	7.6	8.5	9.3	10.1	10.9
30	2.0	3.0	4.0	5.0	5.9	6.8	7.7	8.5	9.3	10.0
31	1.3	2.4	3.3	4.3	5.2	6.1	6.9	7.7	8.5	9.2
32	0.7	1.7	2.7	3.6	4.5	5.4	6.2	7.0	7.7	8.5
33	0.2	1.2	2.1	3.0	3.9	4.7	5.5	6.3	7.0	7.8
34		0.7	1.6	2.5	3.3	4.1	4.9	5.7	6.4	7.1
35		0.2	1.1	1.9	2.8	3.6	4.3	5.1	5.8	6.5
36			0.6	1.5	2.3	3.1	3.8	4.5	5.2	5.9
37			0.2	1.0	1.8	2.6	3.3	4.0	4.7	5.4
38				0.6	1.4	2.2	2.9	3.6	4.2	4.9
39				0.3	1.0	1.7	2.4	3.1	3.8	4.4
40					0.7	1.4	2.1	2.7	3.4	4.0

Table 6.6. (*cont.*)

DT_w	Wet-bulb temperature									
	30	31	32	33	34	35	36	37	38	39
0	100.0	100.0	100.0	100.0	100.0	100.0	100.0	100.0	100.0	100.0
1	92.9	93.0	93.2	93.3	93.4	93.5	93.6	93.7	93.8	93.9
2	86.4	86.6	86.8	87.0	87.2	87.4	87.6	87.8	88.0	88.1
3	80.3	80.6	80.9	81.2	81.5	81.8	82.0	82.3	82.5	82.8
4	74.6	75.0	75.4	75.8	76.1	76.5	76.8	77.1	77.5	77.7
5	69.3	69.8	70.3	70.7	71.2	71.6	72.0	72.3	72.7	73.1
6	64.5	65.0	65.5	66.0	66.5	67.0	67.4	67.9	68.3	68.7
7	59.9	60.5	61.1	61.7	62.2	62.7	63.2	63.7	64.1	64.5
8	55.7	56.4	57.0	57.6	58.1	58.7	59.2	59.7	60.2	60.7
9	51.8	52.5	53.1	53.8	54.4	55.0	55.5	56.1	56.6	57.1
10	48.2	48.9	49.6	50.2	50.9	51.5	52.1	52.6	53.2	53.7
11	44.8	45.5	46.2	46.9	47.6	48.2	48.8	49.4	50.0	50.5
12	41.6	42.4	43.1	43.8	44.5	45.2	45.8	46.4	47.0	47.5
13	38.7	39.5	40.2	40.9	41.6	42.3	42.9	43.6	44.2	44.7
14	35.9	36.7	37.5	38.2	39.0	39.6	40.3	40.9	41.5	42.1
15	33.4	34.2	35.0	35.7	36.4	37.1	37.8	38.4	39.1	39.7
16	31.0	31.8	32.6	33.4	34.1	34.8	35.5	36.1	36.7	37.4
17	28.8	29.6	30.4	31.2	31.9	32.6	33.3	33.9	34.6	35.2
18	26.8	27.6	28.4	29.1	29.9	30.6	31.2	31.9	32.5	33.1
19	24.9	25.7	26.5	27.2	27.9	28.6	29.3	30.0	30.6	31.2
20	23.1	23.9	24.7	25.4	26.1	26.8	27.5	28.2	28.8	29.4
21	21.4	22.2	23.0	23.7	24.5	25.2	25.8	26.5	27.1	27.7
22	19.9	20.7	21.4	22.2	22.9	23.6	24.3	24.9	25.5	26.1
23	18.4	19.2	20.0	20.7	21.4	22.1	22.8	23.4	24.0	24.6
24	17.1	17.8	18.6	19.3	20.0	20.7	21.4	22.0	22.6	23.2
25	15.8	16.6	17.3	18.0	18.7	19.4	20.1	20.7	21.3	21.9
26	14.6	15.4	16.1	16.8	17.5	18.2	18.8	19.5	20.1	20.7
27	13.5	14.3	15.0	15.7	16.4	17.1	17.7	18.3	18.9	19.5
28	12.5	13.3	14.0	14.7	15.3	16.0	16.6	17.2	17.8	18.4
29	11.6	12.3	13.0	13.7	14.3	15.0	15.6	16.2	16.8	17.3
30	10.7	11.4	12.1	12.7	13.4	14.0	14.6	15.2	15.8	16.3
31	9.8	10.5	11.2	11.9	12.5	13.1	13.7	14.3	13.9	15.4
32	9.1	9.8	10.4	11.1	11.7	12.3	12.9	13.4	14.0	14.5
33	8.3	9.0	9.7	10.3	10.9	11.5	12.1	12.6	13.2	13.7
34	7.7	8.3	9.0	9.6	10.2	10.8	11.3	11.9	12.4	12.9
35	7.0	7.7	8.3	8.9	9.5	10.1	10.6	11.2	11.7	12.2
36	6.4	7.1	7.7	8.3	8.9	9.4	10.0	10.5	11.0	11.5
37	5.9	6.5	7.1	7.7	8.3	8.8	9.3	9.9	10.4	10.9
38	5.4	6.0	6.6	7.1	7.7	8.2	8.8	9.3	9.8	10.2
39	4.9	5.5	6.1	6.6	7.2	7.7	8.2	8.7	9.2	9.7
40	4.5	5.0	5.6	6.1	6.7	7.2	7.7	8.2	8.6	9.1

Table 6.7. Percent relative humidity at 85 kPa pressure for various wet-bulb temperatures (°C) and wet-bulb depressions, DT_w (°C). When temperature is below 0°C, saturation vapor pressure is computed for ice.

DT_w	Wet-bulb temperature									
	−10	−9	−8	−7	−6	−5	−4	−3	−2	−1
0	100.0	100.0	100.0	100.0	100.0	100.0	100.0	100.0	100.0	100.0
1	65.9	75.9	77.2	78.4	79.5	80.5	81.5	82.3	83.1	83.9
2	44.4	54.9	57.3	59.5	61.5	63.4	65.1	66.7	68.2	69.5
3	25.6	36.5	39.9	42.9	45.7	48.3	50.6	52.9	54.9	56.8
4	9.4	20.6	24.6	28.4	31.8	35.0	37.9	40.7	43.2	45.5
5		6.8	11.4	15.7	19.7	23.3	26.7	29.9	32.8	35.5
6			4.7	9.1	13.2	16.9	20.4	23.7	26.7
7					4.3	8.3	12.1	15.6	18.9
8						0.8	4.8	8.5	12.0
9								2.3	5.9

	0	1	2	3	4	5	6	7	8	9
0	100.0	100.0	100.0	100.0	100.0	100.0	100.0	100.0	100.0	100.0
1	84.5	85.1	85.7	86.3	86.8	87.3	87.7	88.2	88.5	88.9
2	70.6	71.8	72.9	74.0	74.9	75.8	76.7	77.5	78.2	78.9
3	58.3	60.0	61.5	63.0	64.3	65.6	66.7	67.8	68.9	69.8
4	47.4	49.5	51.4	53.1	54.8	56.3	57.8	59.1	60.4	61.6
5	37.7	40.1	42.3	44.3	46.3	48.1	49.7	51.3	52.8	54.2
6	29.1	31.7	34.2	36.5	38.6	40.6	42.5	44.3	45.9	47.4
7	21.5	24.3	27.0	29.5	31.8	34.0	36.0	37.9	39.7	41.4
8	14.7	17.8	20.6	23.2	25.7	28.0	30.2	32.2	34.1	35.9
9	8.8	11.9	14.9	17.6	20.2	22.6	24.9	27.0	29.0	30.9
10	3.5	6.8	9.8	12.7	15.4	17.9	20.2	22.4	24.5	26.4
11		2.3	5.4	8.3	11.0	13.6	16.0	18.2	20.4	22.4
12			1.4	4.4	7.2	9.8	12.2	14.5	16.7	18.7
13				1.0	3.8	6.4	8.9	11.2	13.4	15.4
14					0.7	3.4	5.8	8.2	10.4	12.4
15						0.7	3.2	5.5	7.7	9.8
16							0.8	3.1	5.3	7.4
17								1.0	3.2	5.2
18									1.2	3.3
19										1.5

Table 6.7. (*cont.*)

DT_w	Wet-bulb temperature									
	10	11	12	13	14	15	16	17	18	19
0	100.0	100.0	100.0	100.0	100.0	100.0	100.0	100.0	100.0	100.0
1	89.2	89.5	89.8	90.1	90.4	90.7	90.9	91.1	91.3	91.5
2	79.4	80.0	80.6	81.2	81.7	82.1	82.6	83.0	83.4	83.8
3	70.6	71.4	72.2	73.0	73.7	74.3	75.0	75.6	76.1	76.6
4	62.5	63.6	64.6	65.5	66.4	67.2	68.0	68.7	69.4	70.1
5	55.3	56.5	57.6	58.7	59.7	60.7	61.6	62.5	63.3	64.1
6	48.7	50.0	51.3	52.5	53.7	54.8	55.8	56.8	57.7	58.5
7	42.7	44.2	45.6	46.9	48.2	49.3	50.4	51.5	52.5	53.4
8	37.3	38.9	40.4	41.8	43.1	44.4	45.6	46.7	47.8	48.8
9	32.4	34.1	35.6	37.1	38.5	39.8	41.1	42.3	43.4	44.5
10	28.0	29.7	31.3	32.9	34.3	35.7	37.0	38.2	39.4	40.5
11	23.9	25.7	27.4	29.0	30.5	31.9	33.2	34.5	35.7	36.9
12	20.3	22.1	23.9	25.5	27.0	28.4	29.8	31.1	32.4	33.5
13	17.0	18.9	20.6	22.3	23.8	25.3	26.7	28.0	29.3	30.4
14	14.1	15.9	17.7	19.3	20.9	22.4	23.8	25.1	26.4	27.6
15	11.4	13.3	15.0	16.7	18.3	19.8	21.2	22.5	23.8	25.0
16	9.0	10.9	12.6	14.3	15.9	17.4	18.0	20.1	21.4	22.6
17	6.8	8.7	10.4	12.1	13.7	15.2	16.6	17.9	19.2	20.4
18	4.9	6.7	8.5	10.1	11.7	13.2	14.6	15.9	17.2	18.4
19	3.1	5.0	6.7	8.3	9.9	11.3	12.7	14.1	15.3	16.5
20	1.6	3.4	5.1	6.7	8.2	9.7	11.0	12.4	13.6	14.8
21	0.2	1.9	3.6	5.2	6.7	8.1	9.5	10.8	12.1	13.2
22		0.6	2.3	3.9	5.4	6.8	8.1	9.4	10.6	11.8
23			1.1	2.7	4.1	5.5	6.8	8.1	9.3	10.5
24			0.0	1.6	3.0	4.4	5.7	6.9	8.1	9.2
25				0.6	2.0	3.3	4.6	5.8	7.0	8.1
26					1.1	2.4	3.6	4.8	6.0	7.1
27					0.2	1.5	2.8	3.9	5.0	6.1
28						0.8	2.0	3.1	4.2	5.2
29							1.2	2.3	3.4	4.4
30							0.6	1.7	2.7	3.7
31								1.0	2.1	3.0
32								0.5	1.5	2.4
33									0.9	1.9
34									0.4	1.3
35										0.9
36										0.4
37										0.1

Table 6.7. (*cont.*)

DT_w	Wet-bulb temperature									
	20	21	22	23	24	25	26	27	28	29
0	100.0	100.0	100.0	100.0	100.0	100.0	100.0	100.0	100.0	100.0
1	91.7	91.9	92.1	92.2	92.4	92.5	92.7	92.8	92.9	93.0
2	84.1	84.4	84.7	85.0	85.3	85.6	85.9	86.1	86.4	86.6
3	77.1	77.5	78.0	78.4	78.8	79.2	79.6	79.9	80.3	80.6
4	70.6	71.2	71.8	72.3	72.8	73.3	73.8	74.2	74.6	75.0
5	64.7	65.4	66.0	66.7	67.3	67.8	68.4	68.9	69.4	69.9
6	59.2	60.0	60.8	61.5	62.1	62.8	63.4	64.0	64.5	65.1
7	54.2	55.1	55.9	56.7	57.4	58.1	58.8	59.4	60.0	60.6
8	49.6	50.5	51.4	52.2	53.0	53.7	54.5	55.2	55.8	56.4
9	45.3	46.3	47.2	48.1	48.9	49.7	50.5	51.2	51.9	52.6
10	41.4	42.4	43.4	44.3	45.2	46.0	46.8	47.6	48.3	49.0
11	37.8	38.8	39.8	40.8	41.7	42.5	43.4	44.2	44.9	45.6
12	34.5	35.5	36.6	37.5	38.5	39.3	40.2	41.0	41.8	42.5
13	31.4	32.5	33.5	34.5	35.5	36.4	37.2	38.1	38.8	39.6
14	28.6	29.7	30.7	31.7	32.7	33.6	34.5	35.3	36.1	36.9
15	26.0	27.1	28.2	29.2	30.1	31.1	31.9	32.8	33.6	34.4
16	23.6	24.7	25.8	26.8	27.8	28.7	29.6	30.4	31.3	32.0
17	21.4	22.5	23.6	24.6	25.6	26.5	27.4	28.2	29.1	29.8
18	19.4	20.5	21.5	22.6	23.5	24.5	25.3	26.2	27.0	27.8
19	17.5	18.6	19.7	20.7	21.6	22.6	23.4	24.3	25.1	25.9
20	15.8	16.9	17.9	18.9	19.9	20.8	21.7	22.5	23.3	24.1
21	14.2	15.3	16.3	17.3	18.3	19.2	20.0	20.9	21.7	22.5
22	12.7	13.8	14.8	15.8	16.7	17.6	18.5	19.3	20.1	20.9
23	11.4	12.4	13.4	14.4	15.3	16.2	17.1	17.9	18.7	19.5
24	10.1	11.2	12.2	13.1	14.0	14.9	15.8	16.6	17.4	18.1
25	9.0	10.0	11.0	11.9	12.3	13.7	14.5	15.3	16.1	16.9
26	7.9	9.0	9.9	10.8	11.7	12.6	13.4	14.2	15.0	15.7
27	7.0	8.0	8.0	9.8	10.7	11.5	12.3	13.1	13.9	14.6
28	6.1	7.1	8.0	8.9	9.7	10.6	11.4	12.1	12.8	13.6
29	5.3	6.2	7.1	8.0	8.8	9.7	10.4	11.2	11.9	12.6
30	4.5	5.4	6.3	7.2	8.0	8.8	9.6	10.3	11.0	11.7
31	3.8	4.7	5.6	6.4	7.3	8.0	8.8	9.5	10.2	10.9
32	3.2	4.1	4.9	5.8	6.5	7.3	8.0	8.7	9.4	10.1
33	2.6	3.5	4.3	5.1	5.9	6.6	7.3	8.0	8.7	9.3
34	2.1	2.9	3.7	4.5	5.3	6.0	6.7	7.4	8.0	8.7
35	1.6	2.4	3.2	4.0	4.7	5.4	6.1	6.8	7.4	8.0
36	1.1	1.9	2.7	3.5	4.2	4.9	5.5	6.2	6.3	7.4
37	0.7	1.5	2.3	3.0	3.7	4.4	5.0	5.7	6.3	6.9
38	0.3	1.1	1.9	2.6	3.2	3.9	4.5	5.2	5.8	6.3
39	0.0	0.8	1.5	2.2	2.8	3.5	4.1	4.7	5.3	5.8
40		0.4	1.1	1.8	2.4	3.1	3.7	4.3	4.8	5.4

Table 6.7. (*cont.*)

DT_w	Wet-bulb temperature									
	30	31	32	33	34	35	36	37	38	39
0	100.0	100.0	100.0	100.0	100.0	100.0	100.0	100.0	100.0	100.0
1	93.1	93.3	93.4	93.5	93.6	93.7	93.8	93.8	93.9	94.0
2	86.8	87.0	87.2	87.4	87.6	87.7	87.9	88.1	88.2	88.4
3	80.9	81.2	81.4	81.7	82.0	82.2	82.5	82.7	82.9	83.1
4	75.4	75.7	76.1	76.4	76.8	77.1	77.4	77.7	78.0	78.2
5	70.3	70.7	71.1	71.5	71.9	72.3	72.6	73.0	73.3	73.6
6	65.5	66.0	66.5	66.9	67.4	67.8	68.2	68.6	68.9	69.3
7	61.1	61.6	62.1	62.6	63.1	63.6	64.0	64.4	64.9	65.3
8	57.0	57.5	58.1	58.6	59.2	59.7	60.1	60.6	61.0	61.5
9	53.1	53.7	54.3	54.9	55.5	56.0	56.5	57.0	57.5	57.9
10	49.6	50.2	50.8	51.4	52.0	52.5	53.1	53.6	54.1	54.6
11	46.2	46.9	47.5	48.2	48.8	49.3	49.9	50.4	50.9	51.4
12	43.1	43.8	44.5	45.1	45.7	46.3	46.9	47.4	48.0	48.5
13	40.2	40.9	41.6	42.3	42.9	43.5	44.1	44.7	45.2	45.7
14	37.5	38.3	39.0	39.6	40.3	40.9	41.5	42.0	42.6	43.1
15	35.0	35.8	36.5	37.1	37.8	38.4	39.0	39.6	40.2	40.7
16	32.7	33.4	34.1	34.8	35.5	36.1	36.7	37.3	37.9	38.4
17	30.5	31.2	31.9	32.6	33.3	33.9	34.5	35.1	35.7	36.3
18	28.5	29.2	29.9	30.6	31.2	31.9	32.5	33.1	33.7	34.2
19	26.6	27.3	28.0	28.7	29.3	30.0	30.6	31.2	31.8	32.3
20	24.8	25.5	26.2	26.9	27.5	28.2	28.8	29.4	30.0	30.5
21	23.1	23.8	24.5	25.2	25.9	26.5	27.1	27.7	28.3	28.8
22	21.6	22.3	23.0	23.6	24.3	24.9	25.5	26.1	26.7	27.2
23	20.1	20.8	21.5	22.2	22.8	23.4	24.0	24.6	25.2	25.7
24	18.7	19.5	20.1	20.8	21.4	22.0	22.6	23.2	23.8	24.3
25	17.5	18.2	18.8	19.5	20.1	20.7	21.3	21.9	22.5	23.0
26	16.3	17.0	17.6	18.3	18.9	19.5	20.1	20.7	21.2	21.7
27	15.2	15.9	16.5	17.1	17.8	18.4	18.9	19.5	20.0	20.6
28	14.1	14.8	15.5	16.1	16.7	17.3	17.8	18.4	18.9	19.5
29	13.2	13.8	14.5	15.1	15.7	16.2	16.8	17.4	17.9	18.4
30	12.3	12.9	13.5	14.1	14.7	15.3	15.8	16.4	16.9	17.4
31	11.4	12.0	12.7	13.3	13.8	14.4	14.9	15.5	16.0	16.5
32	10.6	11.2	11.8	12.4	13.0	13.5	14.1	14.6	15.1	15.6
33	9.9	10.5	11.1	11.5	12.2	12.7	13.3	13.8	14.3	14.7
34	9.2	9.8	10.4	10.9	11.5	12.0	12.5	13.0	13.5	14.0
35	8.5	9.1	9.7	10.2	10.8	11.3	11.8	12.3	12.7	13.2
36	7.9	8.5	9.0	9.6	10.1	10.6	11.1	11.6	12.0	12.5
37	7.3	7.9	8.4	9.0	9.5	10.0	10.5	10.9	11.4	11.8
38	6.8	7.4	7.9	8.4	8.9	9.4	9.9	10.3	10.8	11.2
39	6.3	6.8	7.4	7.9	8.3	8.8	9.3	9.7	10.2	10.6
40	5.8	6.4	6.9	7.4	7.8	8.3	8.8	9.2	9.6	10.0

Table 6.8. Error in relative humidity associated with error in the wet-bulb depression at various temperatures (after Bindon, 1963).

	Temperature (°C)					Error in depression (°C)
	−20	−10	0	+10	+20	
Relative						
humidity	5	3	2	1	1	0.1
error	11	6	4	3	2	0.2

for whatever reason, can generate large errors in relative humidity. The errors depend on the moisture conditions in the atmosphere. Some typical errors are tabulated in Table 6.8 (after Bindon, 1963). Since the wet bulb depression is obtained from two temperature measurements, errors of 0.2°C are not uncommon.

Atmospheric pressure enters into the psychrometric equation, but is not generally measured at research sites. Atmospheric pressure values are frequently obtained from the nearest weather station and adjustments are not always made for differences in elevation. It is obvious from Eq. (6.20) that the difference between saturation and actual vapor pressure is directly proportional to pressure. In the lower atmosphere, pressure decreases about 10 Pa m^{-1} with altitude. An altitude change of 100 m could create a pressure error of 1%. For example, at $P = 100 \text{ kPa}$, $U = 55\%$ and $T_w = 23°C$, the error in vapor pressure would be 4.7 Pa.

The errors associated with radiation, conduction, ventilation, and impure water all tend to raise the wet-bulb temperature. This results in an overestimation of the true vapor content. Radiation error could affect both the dry-bulb and the wet-bulb readings, but it usually warms the wet bulb more and results in a vapor pressure higher than the true value. Care should be taken to insure adequate radiation shielding of the temperature sensors. The Assman psychrometer, Fig. 6.2, is an example of an earlier aspirated (either by spring or electric fan) psychrometer, which minimizes some of the errors. The mercury in glass thermometers are suspended in separate coaxal polished metal tubes to reduce radiation errors.

Conduction of heat to the temperature sensors can affect their performance. This is more pronounced on the wet-bulb sensor that differs more from the temperature of the air. Conduction can occur between the sensor and its supports. Conduction down the signal wires may also be substantial, particularly to the cool wet bulb. Powell (1936) showed the extent to which the conduction error in wet-bulb depression is associated with size of the thermocouple. He defined a factor, a, as the ratio of the observed wet-bulb depression to the true depression,

$$a = \frac{(T_a - T_o)}{(T_a - T_w)}, \tag{6.21}$$

Figure 6.2. The Assman psychrometer.

where T_o is the observed wet-bulb temperature. His results (Table 6.9) demonstrate that conduction is important even with very fine thermocouple wires. This error will be minimized if the wet-bulb wicking extends for at least 1 cm on either side of the junction.

Lourence and Pruitt (1969) evaluated the wet-bulb temperature gradient down the length of a ceramic wick. At least 10 cm of wick was required for the full wet-bulb depression to be reached. Their hollow ceramic wick was 0.95 cm o.d. × 14.6 cm long; it had a pore size of about 0.5 μm.

The wet-bulb temperature is affected by wind movement, especially at lower velocities. It also depends on the size and geometry of the wet-bulb element. Fig. 6.3 illustrates the effect of ventilation on the psychrometric constant for wet bulbs of different size. The ventilation requirement decreases with decreasing size of wet bulb, but the maximum depression is achieved at velocities in excess of 3 m s^{-1}. The WMO (1971) specifies aspiration rates must be between 4 and 11 m s^{-1}.

Table 6.9. The ratio $(T_a - T_o)/(T_a - T_w)$ as a function of thermocouple wire size (after Powell, 1936).

Wire size (mm)	$(T_a - T_o)/(T_a - T_w)$
2.03	0.88
0.457	0.94
0.247	0.96
0.122	0.984

When aspirated wet bulbs are located in shields with axial rather than transverse flow, spoilers should be placed near the intake to increase the turbulence over the wet bulb (Wylie, 1949).

Impurities in the water tend to cause the indicated wet bulb to be higher than the true wet bulb. A saturated solution of NaCl at 20°C on the wet bulb would lower the vapor pressure of water by 25 % (Wylie, 1949). Wicking should be prepared by boiling in a soap solution containing caustic soda and then boiled again in clean water to insure adequate cleanliness.

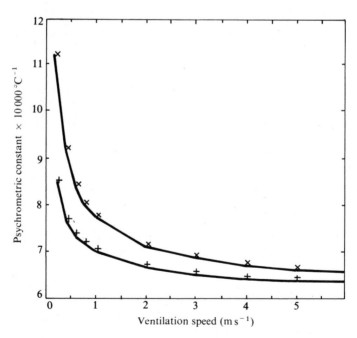

Figure 6.3. Variation of the psychrometric constant with ventilation: ×, spherical bulb 10 mm in diameter; +, cylindrical bulb 8 mm in length and 4 mm in diameter (data of Sworgkin according to Harrison, 1963).

6.2.2   Hair Hygrometer

The hair hygrometer is widely used for routine measurements although it is not usually automated. Hair length varies as a function of its moisture content which is closely associated with the moisture conditions in the surrounding atmosphere. The elongation is greatest for fine blond human hair. The process is nonlinear; the relation between relative humidity and elongation of human hair is shown in Fig. 6.4.

The nonlinearity of this relationship is corrected by the design of the mechanical linkage between the hair and the recorder pen in the hair hygrograph. Hair hygrometers typically have a small negative temperature coefficient that causes the readings to be too high at low temperatures. The temperature coefficient is difficult to eliminate since it is also a function of relative humidity. Middleton and Spilhaus (1953) describe care, maintenance, and calibration procedures for hair hygrometers.

6.2.3   Dew-point Hygrometer

Condensation will occur on a clean surface that is at or slightly below the dew point of the air in contact with it. If the dew-point temperature is measured, the vapor pressure can be obtained from Table 6.1. Therefore, a dew-point hygrometer must be able to cool the surface, observe the occurrence

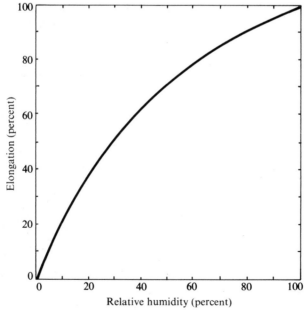

Figure 6.4. Percent elongation of human hair in relation to relative humidity (after Middleton and Spilhaus, 1953).

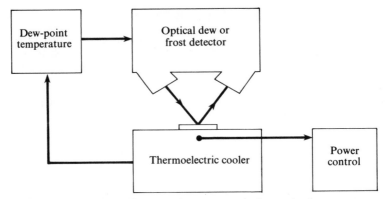

Figure 6.5. A block diagram of an automatic dew-point hygrometer.

of dew, and measure the surface temperatures. These processes can be accomplished manually or automatically.

The dew-point principle is not new, but automation of the principle is rather recent. Most systems use a mirror for the surface and an optical detector to detect the presence of condensate. Moss (1934) utilized a photoelectric cell to detect the formation of dew by observing the change in specular reflection of a light beam as the condensate formed. Thornthwaite and Owen (1940) used a heater circuit connected to a photoelectric cell to raise the temperature of a precooled mirror up to the dew point and measured the temperature with a thermocouple. Initially, mirrors were cooled by conduction, say from a copper rod inserted in a Dewar flask of dry ice (CO_2). Eventually, the Dewar flask was replaced by a thermoelectric cooler using the Peltier effect. One version of this automatic, thermoelectrically-cooled, dew-point hygrometer has been described by Francisco and Beaubien (1963). A block diagram of an automatic dew-point hygrometer is shown in Fig. 6.5.

The mirror surface must receive the condensate in a way that will facilitate its detection, and the mirror temperature must be accurately measured and controlled. The surface must be inert, hydrophobic, reflective, and clean. Generally, rhodium plated silver is used because of the surface hardness and high thermal conductivity. Copper–constantan thermocouples made from 20 μm diameter wire, 5.5 μm diameter bead thermistors, and platinum resistance elements have been used to sense mirror temperatures.

Commercial versions of this instrument quote dew-point accuracies about $\pm 0.5°C$ over a range of -45 to $60°C$ with a response time of $2°C \, s^{-1}$ when the sample flow rate is less than $42 \times 10^{-6} \, m^3 \, s^{-1}$.

6.2.4   Electrical Absorption Hygrometers

Many types of electrical absorption hygrometers have been developed to exploit the effect of moisture content on the electrical properties of various materials. These have been classified by Wexler (1957) into types that depend on the conductivity of aqueous electrolytic solutions, the surface resistivity of impervious solids, and temperature-controlled saturated salt solutions.

6.2.4.1 Dunmore Hygrometer

The Dunmore (1942) hygrometer is probably the best known of the conductivity type. It has undergone many stages of development. Basically, it consists of a polystyrene rod or tube on which a bifilar coil of bare 0.1-mm palladium wire is wound. The element is coated by dipping it into a binder solution of partially hydrolyzed polyvinyl acetate containing a dilute solution of lithium chloride of appropriate concentration. The element is aged for two days at 60% relative humidity and 20°C. The resistance of the bifar winding is a function of its winding geometry and the concentration of lithium chloride. The resistance range of the element can be controlled by fixing the number and the pitch of turns. Sensors of the same resistance value are available for increments of about 16% relative humidity. Eight such sensors connected in parallels cover the 6–99% humidity range. A medium range relative humidity sensor is illustrated in Fig. 6.6. Since lithium chloride is subject to polarization, the resistance between the wires must be measured with an ac resistance meter.

Elements of this type cannot be exposed to an atmosphere of 100% humidity and, thus, are of limited meteorological use. The resistance change ranges over 3 decades; for example, from 1 kΩ to 1 MΩ, and is also a function of the temperature. At 4 m s^{-1} ventilation rate, the time constant is 3 to 6 s. The accuracy of commercial instruments is about $\pm 1.5\%$ relative humidity.

6.2.4.2 Surface Resistivity of Impervious Solids

There are a number of the humidity sensors in this group illustrated by Fig. 6.7. The sensor of Wexler et al. (1955) has a rapid response time, about 0.4 s. A thin film of potassium metaphosphate is deposited on a microscope slide having silver electrodes along its edges. The resistance of the sensor varies

Figure 6.6. A Dunmore-type hygrometer (with protective shield removed) having a medium relative humidity range.

Figure 6.7. An illustration of a humidity sensor with varying surface resistivity on an impervious solid.

from 40 kΩ at 99 % humidity to 10 TΩ at 20 % humidity and is also a function of temperature. The average hysteresis is about 2.7 % relative humidity.

The Jason hygrometer (Cutting et al., 1955) relates the change in either resistance or capacitance of an oxide layer to relative humidity in the atmosphere. The oxide layer is formed on the end of an aluminum rod by anodization in acid electrolyte (17.5 % H_2SO_4, current density of 100 to 1000 A m^{-2} for 30 min). The untreated surface is coated with an insulating layer. A thin porous conducting material (graphite or a thin layer of evaporated metal) is deposited over the oxide layer and onto the insulation. Lead wires are attached to the aluminum rod and the conducting layer.

This transducer can be used for a humidity range from 0 to 100 %, although the sensitivity decreases near saturation. Response time is from 10 to 100 s, being larger at higher humidities. The error is about 3 % relative humidity over a temperature range of −15 to 80°C.

A humidity sensor developed in Finland operates on a capacitance change of a polymer thin film capacitor. A thin gold resistance grid is encapsulated between two polymer layers about one μm thick. The absorption of water vapor by the polymer alters its capacitance. The polymer material is resistant to most chemicals and the calibration is not affected by liquid water.

The humidity range of the sensor is 0 to 100 %. Since the film and metal electrodes are very thin, the instrument responds to a 90 % humidity change in less than one second. The sensitivity is better than ±3 % with hysteresis of less than ±1 %. The sensor has a temperature coefficient of 0.07 % °C^{-1}.

6.2.4.3   Temperature-controlled Saturated Solutions

The type of hygrometer utilizing the principle of temperature-controlled saturated salt solutions is referred to as the "dew cell or dew probe" (Hicks, 1947). A glass wool or dacron wick is wrapped around a tube in which a

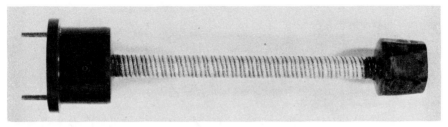

Figure 6.8. A dew-probe-type hygrometer with a bifilar winding on a fiberglass substrate.

temperature sensor is inserted (Fig. 6.8). A bifilar winding of silver or rhodium-plated silver is wound around the cloth. The wick is saturated with a solution of lithium chloride. An alternating current is connected to the bifilar winding through a current limiting resistor. When the wick is wet, the resistance between the wires is low and current will flow causing the element to heat. This in turn will cause water to evaporate, increasing the resistance and reducing the current. Eventually, a temperature equilibrium develops in which the vapor pressure of the lithium chloride is in equilibrium with the vapor pressure of the air. The atmospheric dew point is a function of the lithium chloride dew point or bobbin temperature. This relation is shown in Fig. 6.9. Equations that apply are:

$$T_{dw}(°C) = -23.999\,05 + 0.693\,56\,\text{LiCl}\ T_d(°C),\ \text{LiCl}\ T_d > 12.5°C,$$

$$T_{dw}(°C) = -20.867\,81 + 0.478\,537\,\text{LiCl}\ T_d(°C),\ \text{LiCl}\ T_d < 12.5°C,$$

$$T_{di}(°C) = -18.704\,86 + 0.436\,574\,\text{LiCl}\ T_d(°C),\ \text{LiCl}\ T_d < 12.5°C,$$

$$e(\text{kPa}) = 11.659\,057\,91 + 0.421\,469\,25(\text{LiCl}\ T_d) + 0.011\,280\,44(\text{LiCl}\ T_d)^2$$
$$+ 0.000\,442\,72(\text{LiCl}\ T_d)^3 - 0.000\,000\,31(\text{LiCl}\ T_d)^4$$
$$+ 0.000\,000\,05(\text{LiCl}\ T_d)^5,$$

in which w indicates over water, i indicates over ice, and LiCl T_d is the lithium chloride dew point temperature.

The range of the dew cell is limited from 11 to 100% relative humidity at room temperature. The sensor must be powered at all times to prevent the LiCl salt from absorbing moisture and dripping off. Since LiCl will polarize, ac power is needed. The sensor can be retreated as needed without disturbing the fundamental calibration. The time constant of a dew cell is about 5 min, and the accuracy of dew-point measurement is about ±1.5°C. The operating temperature of the dew cell is considerably above ambient temperature, so care should be taken to minimize heat loss by convection, radiation, and conduction.

The same principle can be applied to individual hygroscopic crystals. If a single crystal such as potassium chloride is subjected to a moist atmosphere, water will be absorbed and a film will form around the crystal. The film will

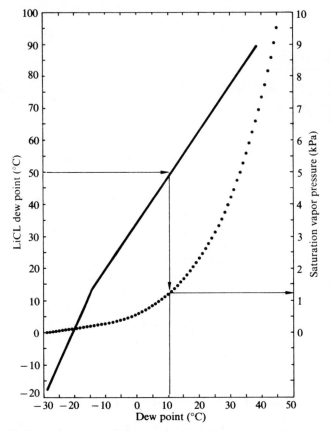

Figure 6.9. Relation between the dew point of lithium chloride or dew-cell bobbin temperature and dew point or vapor pressure of the air. The dotted line is the saturation vapor pressure of water.

continue to grow until equilibrium with the atmosphere is reached. The thickness of the film can be determined by measuring the conductivity across the crystal. If the crystal is in an enclosed chamber, the temperature of the chamber controls the relative humidity, which in turn controls the thickness of the film. In the Wylie hygrometer (Wylie, 1963) the conductivity of the film is used to control the chamber temperature automatically. The equilibrium temperature can be used to derive the corresponding dew-point temperature of the air. The system is capable of determining dew point to about 0.1°C.

6.2.4.4   Carbon Elements

Carbon elements are dimensionally unstable with changes in humidity. The conductivity of this material depends on the presence of carbon particles and as the material expands, the particles separate and the resistance increases.

The carbon element described by Smith and Hoeflich (1954) consists of a polystyrene blank, 100 mm long, 18 mm wide, and 1.2 mm thick, having a thin electrode along either long edge. A mixture of 45% carbon, 32% hydroxyethyl cellulose, 16% polyoxyethylene sorbital, and 7% alkyl aryl polyether alcohol is sprayed on both sides of the blank. This element has low sensitivity between 0 and 40% relative humidity, but the sensitivity increases rapidly above 40%. The calibration tends to go through a maximum above 80% relative humidity, giving two values of relative humidity for the same resistance.

After the element is exposed to the atmosphere for a period of time, the calibration shifts. It also has a hysteresis of about 5% relative humidity. It is, therefore, most useful for one time activity such as radiosonde releases. The 90% response occurs in 30 to 60 s, so the time constant is about 10 to 20 s.

6.2.5   Ion-exchange Hygrometer

When a polystyrene rod is sulfonated with 0.5% silver sulfate and concentrated sulfuric acid at 100°C, the conductivity of the deposit formed is primarily dependent on relative humidity of the air. The calibration curve is also temperature dependent. The thickness of the sulfonated layer, speed of response of the sensor, and the conductivity of the sensor depend on the sulfonation time, which may vary from 1 to 60 min (Pope, 1955).

6.2.6   Infrared Hygrometer

The infrared hygrometer is based on the principle that water vapor absorbs energy at certain wavelengths and not others (Wood, 1963). Consider a test gas flowing through a chamber about 0.25 m in length with an infrared source at one end and lead sulfide detector at the other. The light source is collimated and chopped by a rotating filter wheel transmitting alternately at 2.6 μm and at 2.45 μm. Absorption by water vapor occurs in the longer wavelength, but not in the shorter. The attenuation at the 2.6 μm wavelength, when compared to the attenuation at 2.45 μm, is proportional to the water vapor content of the test gas.

6.2.7   Microwave Hygrometer

The microwave hygrometer utilized two resonant frequency cavities, one filled with dry gas and the other with a test gas (Sargent, 1959). Both cells are subjected to a microwave source of varying frequency. The resonance frequencies of the cells are detected with crystal diodes. The resonance maximum at the diodes will be shifted in proportion to the vapor pressure in the test cell. The time shift is detected electronically.

Tests indicate that this instrument is capable of dew-point accuracies better than 0.2°C in the 0° to 40°C range. The response time of the instrument is limited by the flow rate through the test cell. The time constant is about 10 s.

6.2.8   Thermal Conductivity Bridge

Consider a cell consisting of two concentric cylinders, one heated and the other maintained at constant temperature. If the cylinders are designed to have little radiation or convective heat loss, the thermal conduction between the two is a function of the thermal conductivity of the air. The conductivity of the air in the environmental range is a function of the humidity (Daynes, 1933).

The heated cell and a similar unheated cell both exposed to air are in opposite legs of the bridge. Two other cells with dry gas are located in the other bridge legs. The humidity of the air can be measured over a range of $-40°$ to $50°C$ with a well designed bridge. Since the thermal conductivity of the air is affected by other trace contaminants, the overall uncertainty of dew-point temperature varies from about $0.1°C$ to $1°C$ at low dew points.

6.3   Calibration of Humidity Sensors

Humidity sensors are tedious and difficult to calibrate. Obtaining a source of air with specified temperature and vapor concentration is a major problem. Middleton and Spilhaus (1953) list three primary methods of obtaining different humidities for calibration of hygrometers as

"(1) mixing dry and saturated air;
(2) saturating air at a controlled temperature and then raising its temperature without loss or gain in moisture to any required relative humidity; and
(3) saturating air at a high pressure and then reducing the pressure to a pressure giving the desired relative humidity; or alternatively, because the calibration is usually required at atmospheric pressure the elevated pressure at saturation is selected to give the correct relative humidity when the pressure is reduced to atmospheric."

These methods require facilities and equipment not readily available to most researchers. As an alternative, a saturated salt solution in a closed container will achieve equilibrium with air at a certain relative humidity. The relative humidities are known for different salts and they are relatively independent of ambient temperature. The sensitivity of relative humidity to temperature for solutions of several salts is shown in Fig. 6.10. The temperature coefficients are especially small for solutions of lithium chloride, magnesium chloride, sodium chloride, ammonium sulfate, and potassium sulfate. Specific values of relative humidities observed by Wexler and Hasegawa (1954) are tabulated in Table 6.10. However, their results were obtained under equilibrium conditions. Martin (1965) reported a 4% depression in relative humidity over ammonium nitrate when the vapor was being extracted from the air volume at the rate of $15 \, mg \, h^{-1}$. These results suggest the desirability of closed or recycling systems with large surface area of solutions exposed so equilibrium can be maintained between

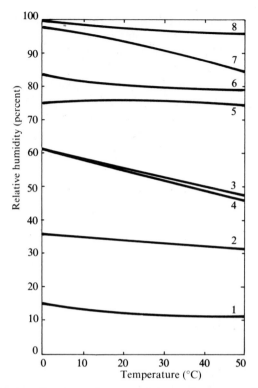

Figure 6.10. Relative humidity of lithium chloride, 1; magnesium chloride, 2; sodium dichromate, 3; magnesium nitrate, 4; sodium chloride, 5; ammonium sulfate, 6; potassium nitrate, 7; and potassium sulfate, 8, as functions of temperature.

Table 6.10. Relative humidity (%) of saturated salt solutions versus temperature in °C. 1, lithium chloride; 2, magnesium chloride; 3, sodium dichromate; 4, magnesium nitrate; 5, sodium chloride; 6, ammonium sulfate; 7, potassium nitrate; and 8, potassium sulfate (Wexler and Hasegawa, 1954).

$T(°C)$	1	2	3	4	5	6	7	8
0	14.7	35.0	60.6	60.6	74.9	83.7	97.6	99.1
5	14.0	34.6	59.3	59.2	75.1	82.6	96.6	98.4
10	13.3	34.2	57.9	57.8	75.2	81.7	95.5	97.9
15	12.8	33.9	56.6	56.3	75.3	81.1	94.4	97.5
20	12.4	33.6	55.2	54.9	75.5	80.6	93.2	97.2
25	12.0	33.2	53.8	53.4	75.8	80.3	92.0	96.9
30	11.3	32.8	52.5	52.0	75.6	80.0	90.7	96.6
35	11.7	32.5	51.2	50.6	75.5	79.8	89.3	96.4
40	11.6	32.1	49.8	49.2	75.4	79.6	87.9	96.2
45	11.5	31.8	48.5	47.7	75.1	79.3	86.5	96.0
50	11.4	31.4	47.1	46.3	74.7	79.1	85.0	95.8

the air and the solution by bubbling air through the solution or by maintaining a large surface area of the solution.

When air is bubbled through salt solutions, neither the salt nor the sensor under test should become contaminated. An inline trap will accomplish this. The test chambers should be protected from heat generated by pumps and fans. When the test chamber is separated from the solution chamber, the temperature of both should be measured to insure that the relative humidity is not altered. Glass containers and connections reduce adsorption of vapor and reduce the equilibrium time of the system.

Literature Cited

Bindon, H. H. (1963). Some fundamental considerations regarding psychrometry. In: Humidity and Moisture. Vol. 3. p. 71–104. Reinhold, New York, 562 pp.

Cutting, C. L., A. C. Jason, and J. L. Wood (1955). A capacitance resistance hygrometer. *J. Sci. Instrum.* **32**:425–431.

Daynes, H. A. (1933). Gas analysis by measurement of thermal conductivity. Cambridge Univ. Press, Cambridge, England.

Dunmore, F. W. (1942). Humidity resistance film hygrometer. U.S. Patent 2,285,421. June 9.

Ferrel, W. (1886). Report on Psychrometric Tables for Use in the Signal Services. Annual Report of the Chief Signal Officer. 1886. Appendix 24, p. 233–259, Washington, D.C.

Francisco, C. and D. J. Beaubien (1963). An automatic dew-point hygrometer with thermoelectric cooling. In: Humidity and Moisture, Vol. 1. p. 165–173. Reinhold, New York. 687 pp.

Goff, Z. A. and Gratch, S. (1946). Low-pressure properties of water from − 160 to 212°F. *Trans. Am. Soc. Heat and Vent. Eng.* **52**:95–122.

Harrison, L. P. (1963). Some fundamental considerations regarding psychrometry. In: Humidity and Moisture, Vol. 3. p. 71–104. Reinhold, New York. 562 pp.

Hicks, W. F. (1947). Humidity measurement by a new system. *Refrig. Eng.* **54**:351–354, 388.

Lourence, F. J. and W. O. Pruitt. (1969). A psychrometer system for micrometeorology profile determinations. *J. Appl. Meteorol.* **8**:492–498.

Martin, S. (1963). The control of conditioning atmosphere by saturated salt solutions. In: Humidity and Moisture. 3:503–507. Reinhold, New York. 562 pp.

Marvin, C. E. (1941). Psychrometric Tables for Obtaining the Vapor Pressure, Relative Humidity, and Temperature of the Dew-point from Readings of the Wet- and Dry-bulb Thermometer. No. 235. U.S. Government Printing Office. Washington.

Middleton, W. E. K. and A. F. Spilhaus (1953). Meteorological Instruments. Univ. Toronto Press, Toronto, Canada. 286 pp.

Moss, E. B. (1934). An apparatus for the determination of the dew point. *Proc. Phys. Soc.* London **45**:450–458.

Pope, M. (1955). Electric hygrometer. U.S. Patent 2,237,006. Dec. 27.

Powell, R. W. (1936). Use of thermocouples for psychrometric purposes. *Proc. Phys. Soc. London.* **48**:406–414.

Sargent, J. (1959). Recording microwave hygrometer. *Rev. Sci. Instrum.* **30**:348–355.

Smith, W. J. and N. J. Hoeflich (1954). The carbon film electric hygrometer element. *Bull. Am. Meteorol. Soc.* **35**:60–62.

Thornthwaite, C. W. and C. J. Owen (1940). A dew-point recorder for measuring atmospheric moisture. *Mon. Weather Rev.* **68**:315–318.

Wexler, A. (1957). Electric hygrometers. *Natl. Bur. Stand. Circ.* **586**.

Wexler (1970). Measurement of humidity in the free atmosphere near the surface of the earth. In: Meteorological Observations and Instrumentation. Meteorol. Monograph 33. p. 262–282. Am. Meteorol. Soc., Boston, Mass. 455 pp.

Wexler, A., S. B. Garfinkel, F. E. Jones, S. Hasegawa, and A. Krinsky (1955). A fast responsing electric hygrometer. *J. Res. Natl. Bur. Stand.* **55**:71–78.

Wexler, A. and S. Hasegawa (1954). Relative humidity temperature relationships of some saturated salt solutions in the temperature range 0° to 50°C. *J. Res. Natl. Bur. Stand.* **53**:19–26.

Wood, R. C. (1963). The infrared hygrometer—Its application to difficult humidity measurement problems. In: Humidity and Moisture. 1:492–504. Reinhold, New York. 687 pp.

World Meteorological Organization. (1971). Guide to Meteorological Instrument and Observing Practice. World Meteorological Organization, Geneva, Switzerland.

Wylie, R. G. (1949). Psychrometry. Report PA-4. CSIRO. Div. of Physics. Sidney, Australia. 55 pp.

Wylie, R. G. (1963). Accurate hygrometry with ionic single crystals. In: Humidity and Moisture. 1:606–616. Reinhold, New York. 687 pp.

Chapter 7

Wind Speed and Direction

7.1 Wind Speed

Environmental assessment of wind speed and direction under field conditions require sensors that are small and rugged yet with low starting speeds and linear response to a wide range of velocities. The WMO, (1971) contains instrument guidelines for many types of routine meteorological measurements, but more sensitive sensors are often needed for air pollution monitoring and environmental measurements. For environmental measurements, the anemometers should generate an electrical signal suitable for transmission and recording. Research needs and wind tunnel studies often require sensitive anemometers with frequency responses as high as 10^5 Hz.

Mazzarella (1973) compared specifications for wind instruments and discussed the need for standardization of test criteria. He defined terms commonly used with anemometers:

The *distance constant* is the distance of air that must pass an anemometer for it to respond to 63% (e.g. $1 - 1/e$) of the step change from the initial to final condition. This term is analogous to the time constant (see Sect. 3.2). It may be determined with an anemometer operating faster or slower than the final condition.

The *threshold* is the speed at which an anemometer starts to operate. With cup or propeller anemometers, it is the speed of air required to start the cups or propeller rotating.

The National Weather Service of the U.S. Department of Commerce, NOAA (Specification No. F460-SP001, dated October 5, 1970) suggests appropriate values for these terms. It suggests that rotating cup anemometers have a starting speed of 270 mm s^{-1}, a maximum distance constant of 1.5 m, and an accuracy of 1% or 70 mm s^{-1}, whichever is greater. The revolutions of the cup assembly are expected to be a linear function of speed and the anemometer is expected to operate over a wide range of speed (22 m s^{-1}).

Although many basic physical phenomena have been employed for indicating the character of wind, the most practical instrument developments of the past have, in general, been motivated by either the dynamic pressure or the thermal cooling action of wind. These two types of action are discussed.

The pressure or force exerted by a wind has been used as a basis for wind speed measurements. Many and varied forms of instruments have been developed and are used today. These instruments fall into three classes: those that measure the dynamic pressure of the wind, those that measure the total dynamic force on an exposed surface, and those that measure the speed or frequency of a pressure pulse. Devices of the first class are limited and are best represented by the common pitot tube. The second class devices are numerous. They include the stationary pressure types such as the pressure plate or the pendulum device and those with moving or rotating pressure areas such as a cup or propeller anemometers. The third class is represented by the vortex trail, and sonic and acoustic Doppler anemometers.

The cooling effect of wind on heated rods and spheres has also been used as an indication of wind speed. Examples are the hot wire, heated thermistor, and heated thermocouple anemometers.

7.1.1   Dynamic Pressure Anemometers

The pressure tube or pitot tube anemometers are used as wind tunnel standards and for special research. The location of the pressure ports are shown in Fig. 7.1. The pressure on the intake port is equal to $P + \frac{1}{2}\rho U^2$, the pressure on the side port is equal to $P - \frac{1}{2}C\rho U^2$, where P is the atmospheric

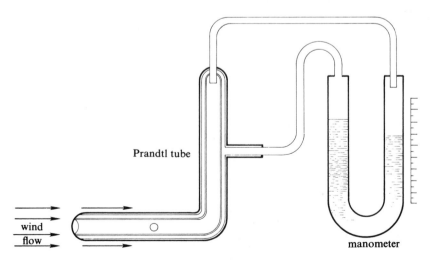

Figure 7.1. Pressure tube anemometer.

pressure, U is the wind speed, and C is a constant less than 1. Subtracting the side port pressure from the intake port pressure results in

$$\Delta P = \frac{\rho U^2(1 + C)}{2}, \tag{7.1}$$

where ΔP is the differential pressure measured with a manometer or differential pressure transducer. The differential pressure is small and difficult to measure under low wind speeds. However, for speeds greater than 5 m s^{-1}, a manometer sensitivity of 0.025 mm of water amounts to an error of less than 1%. Density of the air affects the wind speed readings. Since the pitot tube has to be vaned into the wind, connecting tubing poses a problem for field use. However, these devices are very useful for high speed wind tunnel measurements.

An adaptation of the pressure tube principle has led to a three-dimensional pressure-sphere anemometer that has been used to determine shear stress and eddy fluxes (Thurtell et al., 1970). The pressure sphere contains four additional pressure sensitive intake ports in addition to the main pitot tube, each located 47.5° away from the main tube.

7.1.2   Dynamic Force Anemometers

7.1.2.1   Pendulum Force Measuring Device

The force of the wind acting on an exposed body is small at low wind speeds but may be measurable when a large surface is exposed. An example of this principle, the pendulum anemometer, was described by R. Hook in 1667 and is one of the oldest mechanical anemometers. This anemometer consists of a plate mounted so that it can be vaned perpendicular to the wind. The hanging plate swings about a horizontal axis above the center of gravity but in the plane of the plate. The arc of deflection of the plate caused by the wind force is balanced by its weight. The degree of deflection is calibrated in units of wind velocity. The device flutters in turbulent winds, has a limited nonlinear scale velocity range, and variable sensitivity.

A modified version of the pendulum anemometer, called the Alnor velometer, contains a lightweight flat plate with a restoring force supplied by a hairspring. The degree of deflection of the flat plate is controlled by a stream of air passing through an orifice. The instrument is calibrated to read velocities from 0.1 to 1.2 m s^{-1} with proper selection of orifice size. The directional response makes this unit unsuitable for most field operations.

7.1.2.2   Normal Pressure Plate Anemometer

The normal pressure plate anemometer of Sherlock and Stout (1931) consists of a spring-loaded pressure plate, calibrated to wind velocity. This device differs from the pendulum in that the exposed area is large and is normal to

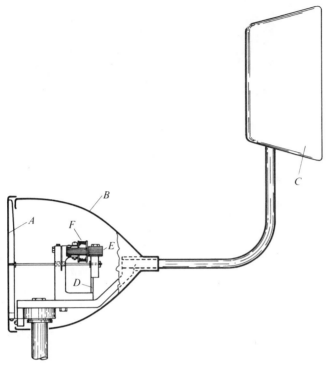

Figure 7.2. The normal-plate anemometer of Sherlock and Stout (from Middleton and Spilhaus, 1953).

the wind. The horizontal displacement is converted to an electrical output signal via a variable transformer. The instrument, shown in Fig. 7.2, consists of the 200 × 230 mm pressure plate, A, which is kept normal to the horizontal wind by a vane C, causing rotation about the ball bearings and slip rings at the top of the supporting mast. The pressure plate is attached at the lower edge to the support frame through a flexible hinge. The plate pressure is conveyed through a push rod to a stiff restoring spring, D. Armature, E, which is attached to the spring, D, varies the electrical impedance of the coil, F, as the air gap or proximity of E varies with the wind pressure. The electrical value is applied to an ac bridge circuit so that the resulting out-of-balance current is an index of the velocity. This instrument was designed for measuring storm gusts with velocities of 10 to 30 m s^{-1}. It could be redesigned for a lower velocity range.

7.1.2.3   Cup Anemometer

The cup-type anemometer is widely used because of its simplicity and sensitivity. It consists of a number (usually three) of cups attached to the ends of horizontal radial arms originating from a hub (Fig. 7.3). The differential

Figure 7.3. Three-cup anemometer (courtesy of Science Associates).

drag force exerted by the wind on the cups varies with the orientation of the cups and causes the wheel assembly to rotate around a vertical shaft.

The speed of rotation is a function of the wind speed. The relation between arm length and cup size was investigated by several workers. Brazier (Middleton and Spilhaus, 1953) concluded that the relation between cup speed and wind speed would be linear when the arm length (distance from the center of the cup to the axis) is equal to the cup diameter. However, commercially available anemometers usually have arm lengths 1.25 times greater than the cup diameter. Patterson (1926) found that the 3-cup anemometer yielded a more uniform torque than the 2- or 4-cup system. The staggered 6-cup system consists of two 3-cup assemblies vertically mounted 60° out of phase and provides more uniform torque than the 3-cup system. Conical-shaped cups were preferred over hemispherical-type cups because they run slower (Marvin, 1934) and the relation between cup speed and wind speed is more linear (Sheppard, 1940). The overrun caused by turbulence is much less with beaded cups than with smooth cups (Marvin, 1934). The peripheral speed of the wheel at the cup centerline is usually 30 to 50% of the wind speed.

Small generators, mechanical or magnetic switches, and photoelectric and capacitance choppers are used to determine the speed of rotation of the cup assembly. The more sensitive anemometers usually have photoelectric choppers that utilize a slotted disc or cylinder attached to the lower end of the cup assembly spindle to interrupt a light beam to a photoconductive cell. The variation in resistance of the cell is used to drive a pulse counter or is converted to a voltage that is proportional to the speed of revolution. The latter type of output is desired for turbulence studies while pulse counting is essential to obtain the resolution needed for profile studies.

Mechanical switches operated by cams or magnetic proximity switches require energy to operate, thereby increasing the threshold. Fritschen and Hinshaw (1972) described the mounting of a small reed switch within an iron filled epoxy cylinder that contained sufficient magnetic attraction to overcome the rotating magnet's attraction for the reed switch. The 2-wire reed switch system provided a positive off–on switch and did not alter the 120 mm s^{-1} threshold.

Miniature dc generators are used to indicate both the speed and direction of rotation of the spindle shaft. These devices provide a low cost output but generally increase the threshold from 100 to 200 mm s^{-1}.

The starting threshold of commercially available anemometers range from a low of 90 mm s^{-1} to a high of 2.24 m s^{-1}. The distance constant, depending on the bearing friction and the mass of the cup assembly, ranges from 0.8 m to 6.4 m (Mazarella, 1973).

7.1.2.4   Vaned Windmill and Propeller Anemometers

This class of anemometer consists of a windmill or propeller mounted on a horizontal spindle. The plane of rotation of the mill or propeller is positioned perpendicular to the wind either manually or with a vane. The pitch of the blades is chosen to cause a rotation about the spindle at a terminal rate which is a linear function of the wind speed. The speed of rotation, at the mean radius, is approximately equal to the wind speed.

A small portable windmill is often used to measure air flow in large ducts. The lightweight blades have low inertia and high torque. Fan rotation is determined from a gear-driven dial. The shroud that protects the rotor makes the device very directional.

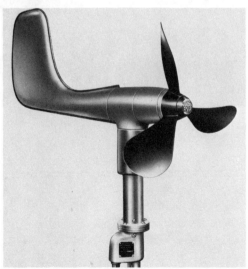

Figure 7.4. The Aerovane anemometer (courtesy of Science Association).

An anemometer with three helicoidal blades on a vane is commercially known as the Aerovane (Fig. 7.4). The propeller drives a small generator that provides an electrical signal for indicator or recorder through a set of slip rings. The direction is transmitted with a pair of synchronous motors. The Aerovane is designed for winds up to 90 m s^{-1}; it has a starting speed of 1.1 m s^{-1} and a distance constant of 4.6 m.

Sensitive propeller anemometers with lightweight 4-bladed rotors have been described by Gill (1975) and MacCready and Jex (1964). These anemometers utilize both generators and photoelectric choppers.

The lightweight propeller anemometer with generator attached has a starting speed of 200 mm s^{-1} and a distance constant of 1 m. The helicodial design causes the response to be nearly cosine, the greatest error being 12% with a 60° angle of attack and may be used in winds up to 30 m s^{-1}. Three propeller anemometers can be mounted in fixed directions (x, y, and z) to obtain the vector components of the wind (Fig. 7.5).

The bivane or vectorvane (Fig. 7.6) is a propeller-type anemometer that is free to rotate around a vertical as well as a horizontal axis. The vector component of the wind is measured directly. Anemometers of this type can respond to wind direction from 0 to 360° and elevation angle up to 60°. The velocity range is 250 mm s^{-1} to 4.5 m s^{-1}.

Figure 7.5. The Gill-type U, V, W anemometer (courtesy of R. M. Young, Co.).

Figure 7.6. A vectorvane (courtesy of R. M. Young, Co.).

7.1.3   Pressure Pulse Frequency Anemometers

7.1.3.1   Vortex Trail Anemometer

When a cylinder is held normal to the wind, eddies are periodically shed from both downstream sides. The eddy shedding frequency is expressed in a dimensionless quantity known as the Strouhal number, S,

$$S = \frac{nd}{U} = f(Re), \tag{7.2}$$

where n is the shedding frequency from one side of the cylinder, d is the cylinder diameter, and U is the stream velocity. The Strouhal number is dependent on the body geometry and roughness, the Reynolds number, the stream turbulence, etc.

Roshcoe (1953), working with a variety of cylinders in a low turbulence wind tunnel, concluded that for a substantial range of selected conditions, S is approximately equal to 0.2 so that $U \simeq 5$ nd. This concept has been utilized in the design of sensitive anemometers consisting of a cylinder equipped with a pulse-sensing device in the form of a hot wire anemometer. The instrument has been used for measurements of velocities down to 0.3 m s^{-1}.

7.1.3.2   Sonic Phasemeter Anemometer

The sonic anemometer relies on the variation of the speed of sound with wind speed (Fig. 7.7). When a sound pressure wave of velocity, c, is impressed on a moving airstream of velocity, U, the wave will travel with the stream at the

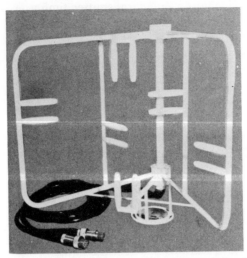

Figure 7.7. Three-dimensional sonic anemometer (courtesy of Weather Measure Corp.).

velocity $(c + U)$ and against the stream at a velocity $(c - U)$. The transmitted sound waves will have an arrival time difference, Δt,

$$\Delta t = \frac{2lU \cos \theta}{(c^2 - U^2)}, \tag{7.3}$$

where l is the acoustic path length, c is the speed of sound in still air, and θ is angle of the wind with respect to the sound wave. Since the distance between transmitter and receiver can be accurately established, only the time difference Δt poses a measuring problem. The value of Δt for wind speeds of 0.3 to 1.5 m s^{-1} is of the order of 0.01 to 1 ms and can be measured with phasemeters.

An arrangement of three sonic anemometers in the x, y, and z directions enables the calculation of the shearing stress and the eddy flux of sensible heat. Design improvements have been presented by Mitsuta (1974) and Kaimal et al. (1974). The sonic anemometer has a resolution of 30 mm s^{-1} over an operating range of 0 to 30 m s^{-1}.

7.1.3.3 Acoustic Doppler Anemometer

Sound waves are scattered in the atmosphere by turbulent eddies. Back scattering of sound is mainly a function of temperature fluctuations in a given volume of air. The amount of energy back scattered is smaller than in any other direction. Back scattered energy is detected by a receiver located with the transmitter. Energy scattered in other directions is referred to as oblique scattering which can be detected by receivers located away from the transmitter. Oblique scattering is a function of wind and temperature eddies.

A typical system includes a sound transmitter, timing device, and frequency tracker. The scattered waves caused by wind induce a Doppler shift, Δf, of the transmitted frequency, f. The wind speed, U, producing the Doppler shift is

$$U = \left[\frac{c}{2} \sin \left(\frac{\theta}{2} \right) \right] \left(\frac{\Delta f}{f} \right), \tag{7.4}$$

where c is the propagation velocity and θ is the angle between the transmitter and receiver.

Beran et al. (1974) discussed a Doppler system for measuring upper air winds. The system has the potential of measuring winds at heights as great as 1 km in height with errors less than 5%.

7.1.4   Thermal Anemometers

The principle of thermal anemometry is not new. Its origin occurred in the early 1900s. Thermal anemometry is based on the measurement of the convective heat loss from a heated sensing element to the surrounding fluid. The rate of heat loss depends on many factors: temperature of the anemometer, geometrical shape, dimensions of the sensor, the fluid velocity, temperature, pressure, density, and thermal properties. If only one of the fluid properties varies, for example, velocity, the heat loss can be interpreted as a measure of that quantity.

The heat transfer, H, from the surface of the heated element at temperature, T_s, to the air, T_a, can be related to the dimensionless Nusselt number, Nu,

$$Nu = \frac{H}{h(T_s - T_a)}, \tag{7.5}$$

where h is the thermal conductivity of the fluid.

In the free convection regime and at low velocities, Nu is a function of the Reynolds number, Re, the Prandtl number, Pr, and the Grashof number, Gr. These dimensionless numbers are defined as

$$Re = \frac{Ud}{v}, \tag{7.6}$$

$$Pr = \frac{v}{k}, \tag{7.7}$$

$$Gr = \frac{g\beta(T_s - T_a)d^3}{U^2}, \tag{7.8}$$

where U is the fluid velocity, d is the diameter of the cylinder, v is the kinematic viscosity, k is the thermal diffusivity of the fluid, g is acceleration due to gravity, and β is the thermal expansion coefficient of the fluid (1/273 for air). For laminar free convection, an expression for cylinders, spheres, and vertical plates is $Nu = 0.54(GrPr)^{1/4}$ (Kreith, 1965).

As the fluid flow changes from free to forced convection, the importance of the Grashof number decreases and the Nusselt number is only a function of Re and Pr ($Nu = Re^n Pr^m$). In the environmental temperature range and for air, Pr is nearly constant ($\simeq 0.71$). Thus, $Nu = 0.66 Re^{1/2} Pr^{1/3}$ applies for flat surfaces in laminar flow (Kreith, 1965).

7.1.4.1   Hot-wire Anemometer

King (1914) developed the fundamental relation for heat transfer from heated wires. His expression for heat loss, H, from a unit length of wire was

$$H = h(T_s - T_a)[1 + (2\pi\rho c_p U/h)^{1/2}] \tag{7.9}$$

where h is the thermal conductivity, ρ is the density, c_p is the specific heat of the air, and U is the wind speed. Combining Eqs. (7.5)–(7.9) leads to

$$Nu = 1 + (2\pi RePr)^{1/2} \tag{7.10}$$

Experiments conducted by Collis and Williams (1959) show that a better approximation for heat transfer in air over the range $0.02 < Re < 44$ is

$$Nu = \left[\frac{(T_s + T_a)}{2T_a}\right]^{0.17} (0.24 + 0.56 Re^{0.45}). \tag{7.11}$$

When the length of wire, l, is short so that heat is transferred to the mounting studs, individual calibration of the sensors is required. At equilibrium, the total heat loss from the sensor of diameter d is related to the power consumed, P;

$$H\pi dl = P = i^2 R. \tag{7.12}$$

For a hot-wire anemometer at operating-temperature resistance, R, where the resistance at ambient temperature is R_a, the calibration factor takes the form

$$\frac{i^2 R}{(R - R_a)} = A + BU^n. \tag{7.13}$$

Constants A, B, and n are determined experimentally.

The sensing element of a hot-wire anemometer is constructed from 0.013 to 0.13-mm wire. The selection of wire type is a compromise of ruggedness, thermal, chemical, and electrical stability. Hence, platinum is most frequently used. The sensor is usually operated in a range of 200° to 500°C. Its useful operating range is from a few mm s^{-1} to supersonic speeds. Since the sensor has little mass, it can respond to changes in fluid velocity at frequencies up to 500 Hz. A hot-wire anemometer is fragile and easily contaminated, therefore, its chief use is in wind tunnel studies.

In recent years, hot-film anemometers have replaced hot-wire anemometers because they have a faster response, approaching 1 MHz. The hot-film sensing element is essentially a platinum film on the surface of a quartz rod or substrate. The hot films are less susceptible to fouling, have lower conduction to the support, better frequency response, and offer more flexibility in sensor configuration.

There are two modes of operation of the hot-wire anemometer—constant current and constant temperature. In the constant current mode, a current is supplied through a large current limiting resistor to the anemometer located in a bridge configuration. As the heat transfer to the environment increases, due to increased wind speed, the temperature and resistance of the anemometer will decrease. The resistance decrease will result in a voltage inbalance of the bridge which when amplified with a nonlinear amplifier is an indication of the wind speed.

In the constant temperature mode, the anemometer is also located in a bridge configuration. As the speed past the sensor increases, the resulting cooling will decrease the resistance. The decrease in resistance causes a decrease in the bridge output which when amplified causes an increase in current to the sensor. If the amplifier has sufficient gain, it will tend to keep its inputs very close to balance. When solid state circuitry is used, the constant temperature system is preferred because (1) it is compatible with film-type sensors, (2) it prevents sensor burnout that may occur with a constant current system during a sudden lull in wind, (3) linearization is possible, (4) it can be temperature compensated, and (5) it yields a direct dc output.

7.1.4.2   Thermistor Anemometer

In principle, the thermistor anemometer is very similar to the hot-wire anemometer. The major differences are that the thermistor is maintained at a lower temperature and the self-heating characteristics of the thermistor are utilized. At zero current, the resistance of a thermistor decreases approximately logarithmically with increasing temperature. When current is applied, the thermistor will self-heat. As shown in Sect. 3.3.2.2, the self-heating is accentuated by the temperature resistance relationship.

The thermistor is placed in a bridge circuit, and the current through it is controlled to maintain the thermistor at a constant resistance or temperature, for example, 120°C. The current required is a function of the cooling due to convection, and the relation takes the form of Eq. (7.13).

Thermistor anemometers, although lacking the frequency response of hot wires, have a greater sensitivity and are more rugged. Bead thermistors located on the head of a bivane arrangement have been used to study the fine structure of the atmosphere (Cramer and Record, 1953).

7.1.4.3   Heated Thermocouple Anemometer

Hukill (1941) described a thermocouple anemometer designed to measure air flow in refrigerator railroad cars. The anemometer consisted of a single-junction thermopile, one junction being heated by a constant dc current

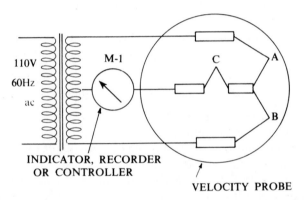

Figure 7.8. Hastings–Raydist heated-thermopile anemometer. A and B are heated, and C is the nonheated thermocouple junction (courtesy of Teledyne Hastings–Raydist).

heater and the other used to compensate for ambient temperature. In contrast to hot-wire and heated thermistors, the thermocouple is heated about 25°C above ambient. A nondirectional version of the thermocouple anemometer was described by Fritschen and Shaw (1961).

Hastings (1948) improved the frequency response of the thermocouple anemometer by supplying an ac heating current directly to a single-junction thermopile (Fig. 7.8). The dc voltage between the center point of the thermopile and another thermocouple in opposition for temperature compensation is an indication of the wind speed. The calibration is logarithmic, being very sensitive to low wind speeds and decreasing in sensitivity at 1 m s^{-1}. A fast response version of this anemometer was described by Gill (1954) for micrometeorological investigations.

7.1.5   Anemometer Calibration

Anemometers are calibrated in wind tunnels, on turn tables, in drag tunnels, or by out-of-door comparisons. When wind tunnels are used for calibration purposes, the test cross section should be at least twice as large as the anemometer to be calibrated. With smaller tunnels, acceleration of air around the sensor and edge effects can cause errors. The tunnel should also be able to operate at steady speeds for at least two minutes.

Standards used in wind tunnels include the pressure tube, the vortex trail anemometer, or another certified anemometer preferably of the same design as the instrument to be calibrated. The range of operations of these standards are different. The pitot tube is not very sensitive at low wind speeds because the differential pressure is small. The manometer has a sensitivity of ± 0.025 mm of water which amounts to an error in speed readings of less than 1 % for tunnel speeds greater than 5 m s^{-1}. The error increases to approximately 10 % at 1.3 m s^{-1}

The eddy shedding hot wire can be used to measure air speeds from about 200 mm s^{-1} to 5 m s^{-1}. The frequency at which the eddies are shed depends on the air velocity, wire diameter, and viscosity according to Eq. (7.2).

The controls on wind tunnels generally are not effective below 200 mm s^{-1}. Drag tunnels are used for calibration in this range. The anemometer is mounted on a trolley and is moved at a constant speed across a known distance. Such a system can be very accurate; however, it must be in thermal equilibrium.

Turn tables can be used for calibration. Since the rotation of the turn table causes some air to be in motion, it cannot be considered an absolute method. However, very good results are achieved when anemometers are compared with certified anemometers of the same design.

If the above facilities are not available, anemometers may be compared against each other on a horizontal bar out-of-doors. The bar should be located away from any obstructions and perpendicular to the mean wind direction. The anemometers are mounted far enough apart (at least two cup diameters) to prevent mutual interference. Frequent interchange of instrument location will insure the lack of bias due to location. This is a good procedure for comparing anemometers to be used in wind profile studies. However, it has been the experience of the authors that calibration differences of anemometers of the same design and with clean bearings are due to cup irregularities. When cup assemblies were made with close tolerances, no differences could be found in their calibration factors.

7.1.6   Anemometer Location

The WMO (1971) specifies that anemometers for climatological measurements be exposed 10 meters above open terrain. Open terrain is defined as 10 times the height away from the obstruction. If such a location is not available, the anemometer should be located away from obstructions and at a height to represent the unobstructed wind speed at the 10 meter height.

Mounting anemometers on the roofs of buildings or on towers may affect the indicated wind speed. Moses and Daubek (1961) studied the effect of a tower (cross section of 2 m^2) on the wind speed. They reported that air flow on the lee side of the tower may be reduced to one-half of its true value in light winds and nearly 25% for 5 m s^{-1} wind speed. An increase of 30% was observed when the wind blowing toward the anemometer made an angle of 20° to 40° with respect to the sides of the tower adjacent to the anemometer.

When wind speeds are measured for surface roughness or flux calculations, additional restrictions are imposed on location and mounting materials. For wind profile measurements, the lowest anemometer should be located at a height of at least five times the roughness height. The highest anemometer should have an upwind fetch of 50 to 100 times the height of measurement above the surface (Tanner, 1963).

Anemometers should be located as far as possible from any mast or support. The mounting should be rigid enough to eliminate vibration and also maintain the anemometer in a vertical position. The direction of cup rotation with respect to wind direction to the mounting apparatus and nonsymmetrical anemometers will affect the indicated wind speed (Rider, 1960). Vertical spacing also affects the indicated speed. Wind speeds were found to be 2% greater when cup-type anemometers were stacked 200 mm apart (Fritschen, 1967).

7.2   Wind Direction

Wind direction is a parameter of only indirect interest in many environmental studies. Its measurement is often omitted, despite the usefulness of wind direction in explaining biological phenomena.

The direction of the wind may be important at either the mesoscale or the microscale. Mesoscale movements include local winds, as well as air mass movements. In hilly areas, for example, the local winds move upslope during periods of daytime heating, and downslope during nighttime cooling. Wind direction becomes important on the microscale in the turbulence and flow studies associated with air pollution. The accuracies required for direction measurements in mesoscale phenomena are rather low. Higher accuracies, perhaps an order of magnitude higher, are needed for measurements of wind direction associated with microscale phenomena.

Wind direction has been measured for over two thousand years with various types and shapes of devices. However, the response of the basic wind vane has undergone a critical review and analysis only within the last decade (MacCready and Jex, 1964; Wieringa, 1967). Mazarella (1972) has summarized terms and definitions applicable to wind vanes in an attempt to standardize usage.

In a steady wind, a wind vane will indicate the true direction. When the wind suddenly changes direction, the vane response depends on the dynamic characteristics of its design. It will respond with an exponential approach to the new direction only if it is critically damped. A distance constant (see Sect. 3.2) can be defined for a critically damped vane, and the time required to respond to the step change can be quite large. The response of damped and critically damped wind vanes is illustrated in Fig. 7.9. In practice, wind vanes are slightly underdamped and tend to oscillate about the true wind direction, as illustrated by line 3.

The equations for vane motion are given here to illustrate descriptive terms. According to Wieringa (1967), the vane torque, T, per unit angle is

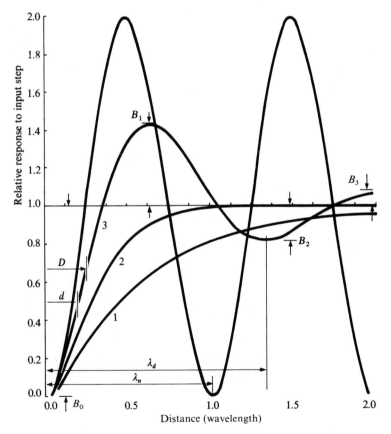

Figure 7.9. The relative response to a step input of an overdamped (1), critically damped (2), and damped (3) wind vane, where λ_n is the natural wavelength, λ_d is the damped wavelength, D is the distance constant, d is the delay distance, and the B's are the amplitudes of the damped oscillation.

produced by a force, F, acting on the vane with centroid radius r,

$$T = \frac{rF}{\theta}, \tag{7.15}$$

where θ is the angle between the wind and the vane. When the vane is moving, an additional force resulting from the fin speed, $rd\theta/dt$, is acting on the vane. This action changes the attack angle of the vane θ_v. The vane motion is described by

$$\theta_v = \tan^{-1}\left[\frac{(U\sin\theta + r\,d\theta/dt)}{U\cos\theta}\right]. \tag{7.16}$$

The moment of inertia, I, is related to the torque [Eq. (7.15)] by

$$-I\frac{d^2\theta}{dt^2} = rF = T\theta_v = T\theta + \left[\frac{rT(d\theta/dt)}{U}\right]. \tag{7.17}$$

The term (rT/U) is defined as the aerodynamic damping of the vane M.

Assuming constant torque and an exponential relation between variables θ and t,

$$\theta = C\exp bt, \tag{7.18}$$

where C and b are constants, we have

$$\frac{d\theta}{dt} = bC\exp bt \quad \text{and} \quad \frac{d^2\theta}{dt^2} = b^2C\exp bt. \tag{7.19}$$

Substitution of Eqs. (7.18) and (7.19) into Eq. (7.17) yields the general form

$$Ib^2 + Mb + T = 0, \tag{7.20}$$

with solution

$$b = -\frac{M}{2I} \pm \left[\left(\frac{M^2}{4I^2}\right) - \left(\frac{T}{I}\right)\right]^{1/2} = -f \pm g. \tag{7.21}$$

The displacement of the vane at anytime is

$$\theta_v = [\exp(-ft)][J\exp gt - K\exp(-gt)], \tag{7.22}$$

where J and K are constants of integration required to satisfy the equation. From Eq. (7.21), the vane is said to be overdamped (Fig. 7.9, line 1) when

$$\left(\frac{M^2}{4I^2}\right) > \left(\frac{T}{I}\right), \tag{7.23}$$

critically damped (Fig. 7.9, line 2) when

$$\left(\frac{M^2}{4I^2}\right) = \left(\frac{T}{I}\right). \tag{7.24}$$

and underdamped (Fig. 7.9, line 3) when

$$\left(\frac{M^2}{4I^2}\right) < \left(\frac{T}{I}\right).$$ (7.25)

The *damping ratio h* is obtained from the ratio of a swing amplitude B_2 to the previous swing amplitude on the same side, B_0,

$$h \propto \frac{B_2}{B_0} = \frac{B_3}{B_1}.$$ (7.26)

Mathematically, the damping ratio can be expressed as

$$\frac{B_2}{B_0} = \exp\left[\frac{-2\pi h}{(1 - h^2)^{1/2}}\right]$$ (7.27)

When a second oscillation is absent, the damping is defined by

$$\frac{B_1}{B_0} = \exp\left[\frac{-\pi h}{(1 - h^2)^{1/2}}\right]$$ (7.28)

or

$$h = \left[1 + \left(\frac{\pi}{\ln(B_0/B_1)}\right)^2\right]^{1/2}$$ (7.29)

The *delay distance, d,* is the product of the wind speed and the time required for the vane in a wind tunnel to move from the point of release to 50% of the distance to final direction (Fig. 7.9).

Damped wavelength, λ_d, is the product of the wind tunnel speed and the time for one complete oscillation.

Undamped wavelength, λ_n, is computed from the damped wavelength and the damping ratio by

$$\lambda_n = \lambda_d(1 - h^2)^{1/2},$$ (7.30)

or from the delay distance and the damping ratio by

$$\lambda_n = d(6.0 - 2.4h).$$ (7.31)

It is recommended that λ_n be computed as the average of Eqs. (7.30) and (7.31).

Related to the delay distance is the *distance constant*. It is the distance of air passing the vane prior to 63% recovery of a·step change. The distance constant D can be computed as

$$D = \frac{r}{2h^2}.$$ (7.32)

Maximum overshoot with sinusoidal fluctuations is obtained from the relation of the amplitude ratio (undamped wavelength/impressed gust wavelength) to the gust wavelength.

Attainable and satisfactory characteristics for wind-direction vanes (WMO, 1971) are a linearity and resolution of $\pm 2°$ to $\pm 5°$, a damping ratio of 0.3 to 0.7, and a range of 0.5 to 50 m s^{-1}. The National Weather Service specifications include a maximum distance constant of 1.2 m, a damping ratio of 0.4 for a 10° initial attack angle, a threshold of 310 mm s^{-1} at 10°, and an accuracy of $\pm 3°$.

A critical review and evaluation of wind vane design was accomplished by Wieringa (1967). He concluded that the single-wedge fin (splayed) and the streamlined fin are inferior to the simple flat fin in all respects. A vane quality factor, K_v, was used for evaluation,

$$K_v = \frac{a}{[(m/S)(1 + r_w/r)]}, \qquad (7.33)$$

where $a = 4.9A/(EA + 2)$; A, the aspect ratio $= s^2/S$; s, the vertical span of the air foil; S, area of air foil; E, edge correction = semiperimeter/wing span; m, weight of vane; r_w, radius of movement of counter weight (centroid); and r, radius of movement of the vane (lying approximately 1/4 of vane width from leading edge). Using the vane quality factor, overshoot, and damped wavelength, he designed and tested new vanes for macro- and micrometeorological purposes. Specifications for these vanes are given in Table 7.1.

The presence of the factor $1/(1 + r_w/r)$ and the absence of the counter weight mass in Eq. (7.33) means that a heavy counterweight on a short arm must be used. A light counterweight on a long arm increases the moment of inertia. Due to the fact that r_w is small, it is not worth the trouble to shape the counterweight.

Specifications are not often available for commercially available macrometeorological vanes. Limited specifications are usually given for the micrometeorological vanes. However, the definitions make it difficult to compare the performance of vanes from different manufacturers.

Table 7.1. Specifications of macro- and micrometeorological vanes (Wieringa, 1967).

	Macrometeorological vane	Micrometeorological vane
Fin	1.0-mm flat aluminum	0.2-mm ring shaped
A	3.2	5.0
r	0.45 m	0.235 m
r_w	0.15 m	0.14 m
K_v (calculated)	0.76	2.60
K_v (experimental)	0.90	2.8
Damping ratio	0.30	0.54
Damped wavelength	7.0 m	4.6 m

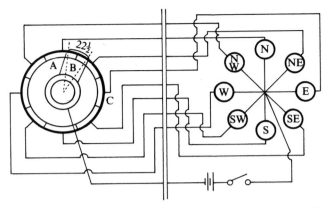

Figure 7.10. Sixteen-point direction indicator (from Middleton and Spilhaus, 1953).

Wind vane direction has been indicated by mechanical and electrical means. Electrical means include (1) lights operated by a contact, (2) a variable voltage divider, (3) a pair of position motors, and (4) a shaft encoder.

A 16-position light system is illustrated in Fig. 7.10. The wiper A is attached to the shaft of the wind vane and to the spring-loaded contact B. The contact, B, completes the circuit by contacting one or two of the octants, C. Lamps or relays connected to marking pens are acuated by the completed circuit. It is possible to determine the wind direction to the nearest $22\frac{1}{2}°$ with this procedure.

When the wiper shaft of a low-torque single-turn potentiometer is connected to the wind vane shaft, the potentiometer can be used as a variable voltage divider. If a constant voltage is impressed on the fixed portion of the potentiometer, the voltage drop across the variable portion is an indication of the wind direction. If a 2-wire system is used, as shown in Fig. 7.11, consideration should be given to the voltage drop of the lead wires, and an additional resistor of the same value as the potentiometer is required for calibration purposes. This calibration resistor is not required with a 3-wire system.

Since a single-turn potentiometer has an approximate 6° electrical gap, the wind direction has an uncertainty of $±3°$. To reduce this uncertainty,

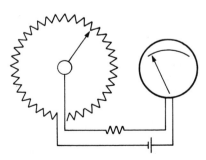

Figure 7.11. Potentiometer-type indicator (from Middleton and Spilhaus, 1953).

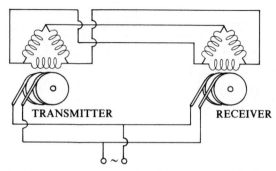

Figure 7.12. Selsyn-type indicator (from Middleton and Spilhaus, 1953).

special dual-wiper potentiometers with a solid state switching network have been devised to indicate the direction from 0° to 540°. This design also minimizes problems that occur when the direction is recorded on a chart. If the direction varies about the dead band, the output from the single potentiometer varies between zero and full-scale, and excessive pen travel can rapidly wear through the chart paper. This will not occur with the dual-wiper potentiometer circuit.

Some wind vanes utilize a pair of dc powered Selysn transmitters with torodial resistance coil having three equally spaced taps (Fig. 7.12). The vane is attached to the shaft having two brushes spaced 180° apart. The voltage differentials are transmitted to the receiver indicator, causing it to rotate in tandem with the transmitter.

Shaft encoders have been attached to the shaft of wind vanes to indicate wind direction. The encoder consists of a multiband disc with 2, 4, 8, etc. patterns on the bands. Either wipers or photocells are used to sense the off/on conditions of each of the bands. These conditions are translated into wind direction to the nearest degree. Although this system is very good, it is more expensive than the popular potentiometer design.

Literature Cited

Beran, D. W., B. C. Willmarth, F. C. Carsey, and F. F. Hall (1974). An acoustic Doppler wind measuring system. *J. Acoust. Soc. Am.* **55**:334–338.

Collis, D. C. and M. J. Williams (1959). Two-dimensional convection from heated wires at low Reynolds numbers. *J. Fluid Mech.* **6**:357–384.

Cramer, H. E. and F. A. Record (1953). The variation with height of the vertical flux of heat and momentum. *J. Meteorol.* **10**:219–226.

Fritschen, L. J. and R. H. Shaw (1961). A thermocouple-type anemometer and its use. *Bull. Am. Meteorol. Soc.* **42**:42–46.

Fritschen, L. J. (1967). A sensitive cup-type anemometer. *J. Appl. Meteorol.* **6**:695–698.

Fritschen, L. J. and R. Hinshaw (1972). A reed switch anemometer. *J. Appl. Meteorol.* **11**:742–744.

Gill, G. C. (1954). A fast response anemometer for micrometeorological investigations. *Bull. Am. Meteorol. Soc.* **35**:69–75.

Gill, G. C. (1975). Development and use of the Gill UVW anemometer. *Boundary-Layer Meteorol.* **8**:475–495.

Hastings, C. E. (1948). A new type of instrument for measuring air velocity. *Am. Inst. Elec. Eng. Misc. Paper* **49**:23.

Hukill, W. Y. (1941). Characteristics of thermocouple anemometers. In: Temperature, Its Measurement and Control in Science and Industry. p. 666–672. Reinhold, New York.

Kaimal, J. C., J. T. Newman, A. Bisberg, and K. Cole (1974). An improved three-component sonic anemometer for investigation of atmospheric turbulence. In: Flow: Its Measurement and Control in Science and Industry, p. 349–359. Instrument Soc. Amer. Vol. 1. 1048 pp.

King, L. V. (1914). On the convection of heat from small cylinders in a stream of fluid. *Philos. Trans. R. Soc. London Ser.* A **214**:373–432.

Kreith, F. (1965). Principles of heat transfer. International Textbook Co., Scranton, Pa. 665 pp.

MacCready, P. B. and H. R. Jex (1964). Response characteristics and meteorological utilization of wind vane sensors. *J. Appl. Meteorol.* **3**:182–193.

Marvin, C. F. (1934). Recent advances in anemometry. *Mon. Weather Rev.* **62**:115–120.

Mazzarella, D. A. (1972). An inventory of specifications for wind measuring instruments. *Bull. Am. Meteorol. Soc.* **53**:860–871.

Middleton, W. E. K. and A. F. Spilhaus (1953). Meterological Instruments. Univ. Toronto Press, Toronto, Canada, 286 pp.

Mitsuta, Y. (1974). Sonic anemometer-thermometer for atmospheric turbulence measurements. p. 341–347. In: Flow: Its Measurement and Control in Science and Industry. Instrument Soc. Amer. Vol. 1. 1048 pp.

Moses, H. and H. G. Daubek (1961). Errors in wind measurements associated with tower-mounted anemometers. *Bull. Am. Meteorol. Soc.* **42**:190–194.

Patterson, J. (1926). The cup anemometer. *Trans. R. Soc. Can. Sect.* 3, **20**:1–54.

Rider, N. E. (1960). On the performance of sensitive cup anemometers. *Meteorol. Mag.* **89**:209–215.

Roshcoe, A. (1953). On the development of turbulent waves from vortex streets. National Advisory Committee for Aeronautics Technical Note No. 2913, March.

Sheppard, P. A. (1940). An improved design of cup anemometer. *J. Sci. Instrum.* **17**:218–221.

Sherlock, R. H. and M. B. Stout (1931). An Anemometer for a Study of Wind Gusts. Univ. of Michigan Res. Bull. No. 20.

Tanner, C. B. (1963). Basic Instrumentation and Measurements for Plant Environment and Micrometeorology. Dept. Soils Bull. 6. Univ. of Wisconsin, Madison, Wisc.

Thurtell, G. W., C. B. Tanner, and M. L. Wesely (1970). Three-dimensional pressure-sphere anemometer system. *J. Appl. Meteorol.* **3**:379–385.

Wieringa, J. (1967). Evaluation and design of wind vanes. *J. Appl. Meteorol.* **6**:1114–1122.

World Meteorological Organization (1971). Guide to Meteorological Instrument and Observing Practices. World Meteorological Organization, Geneva, Switzerland.

Chapter 8

Pressure

8.1   Introduction

The pressure of a fluid at rest is equal to the force exerted perpendicularly by the fluid on a unit area of any bounding surface,

$$P = \frac{F}{A},$$
(8.1)

where P is the pressure, F is the force, and A is the area. SI units are pascals ($1 \text{ Pa} = 1 \text{ N m}^{-2} = \text{kg m}^{-1} \text{ s}^{-2}$). The millibar is a unit of pressure generally used by meteorologists by international agreement. Under standard conditions $1 \text{ atm} = 1013.25 \text{ mb} = 101.325 \text{ kPa} = 760 \text{ mm Hg}$.

A change in pressure is given by the change in the head of fluid, h, with density, ρ, and acceleration of gravity g,

$$dP = \rho g \, dh.$$
(8.2)

Pressure is unaffected by the shape of the confining vessel and is transmitted undiminished throughout the fluid to all of the bounding surfaces.

The manometer was used by Boyle as early as 1662 for precise determination of pressure. The manometer consists of a transparent tube usually in the form of a U and partially filled with one or more liquids. When pressure is applied to one leg of the manometer, the magnitude of the applied pressure is indicated by the change in head, or a difference between the heights of the liquid in the two legs, in accord with Eq. 8.2. The manometer is normally filled with either water or mercury, though other liquids can be used. With water, and under steady state conditions, the uncertainty is about 0.2% or less.

According to Benedict (1977), when more than one liquid is involved, the pressures (Fig. 8.1) are

$$P_u + \rho_1 g(h_1 + h_2 + \Delta h) = P_r + \rho_2 g h_2 + \rho_m g\, \Delta h, \qquad (8.3)$$

or

$$P_u - P_r = \Delta P = \rho_m g\, \Delta h \left[1 + \frac{\rho_2}{\rho_m} \frac{h_2}{\Delta h} - \frac{\rho_1}{\rho_m}\left(\frac{h_1 + h_2}{\Delta h} + 1\right)\right]. \qquad (8.4)$$

If $\rho_1 = \rho_2$, then

$$\Delta P = \rho_m g\, \Delta h\left[1 - \frac{\rho_1}{\rho_m}\left(\frac{h_1}{\Delta h} + 1\right)\right], \qquad (8.5)$$

and if $h_1 = 0$, then

$$\Delta P = \rho_m g\, \Delta h\left(1 - \frac{\rho_1}{\rho_m}\right). \qquad (8.6)$$

The subscripts are m, the manometer fluid; u, unknown; and r, reference.

The densities used in the above equations are for standard temperature and pressure, and corrections must be applied for other conditions. The density of water and mercury at various temperatures are given in Table 8.1. Variations in the density of water with temperature are given by

$$\rho_w = 999.895 + 50.692 \times 10^{-3}T - 7.278 \times 10^{-3}T^2 + 31 \times 10^{-6}T^3, \qquad (8.7)$$

and for mercury by

$$\rho_{Hg} = 13\,595.089 - 2.467T + 352 \times 10^{-6}T^2. \qquad (8.8)$$

Corrected densities should be used in Eqs. (8.4), (8.5), and (8.6).

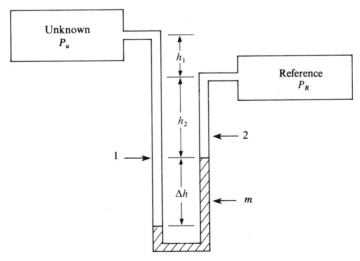

Figure 8.1. The manometer containing the manometer fluid (m) and two other fluids (1 and 2), unknown pressure (P_u), reference pressure (P_r), and liquid heights (h).

Table 8.1. The density of water and mercury at various temperatures.

Temperature (°C)	Water (kg m^{-3})	Mercury[a] (kg m^{-3})
0	999.87	13 595.089
5	999.99	13 582.764
10	999.73	13 570.457
15	999.13	13 558.166
20	998.23	13 545.892
25	997.07	13 533.635
30	995.67	13 521.393
35	994.06	13 509.167
40	992.24	13 496.956
45	990.25	13 484.760
50	988.07	13 472.579
55	985.73	13 460.412

[a] Based upon the value of 13 545.892 kg m^{-3} at 20°C.

Table 8.2. Corrections for gravity at sea level for various latitudes.

Latitude (degrees)	g_ϕ (m s^{-2})
0	9.780 36
5	9.780 75
10	9.781 91
15	9.783 81
20	9.786 38
25	9.789 56
30	9.793 24
35	9.797 32
40	9.801 67
45	9.806 16
50	9.810 65
55	9.815 01
60	9.819 11
65	9.822 81
70	9.826 01
75	9.828 60
80	9.830 51
85	9.831 68
90	9.832 08

In addition to the temperature corrections, manometer readings are also corrected for gravity and elevation. The acceleration due to gravity at any latitude, ϕ, is given by

$$g_\phi = 9.806\,16(1 - 2.637 \times 10^{-3} \cos 2\phi + 5.9 \times 10^{-6} \cos^2 \phi). \quad (8.9)$$

Acceleration due to gravity at various latitudes are given in Table 8.2. Elevation correction is given by

$$g = g_\phi - 3.086 \times 10^{-6}h + 1.118 \times 10^{-6}(h - h'), \quad (8.10)$$

where h is the elevation of the instrument above sea level in meters and h' is the elevation of the general terrain for a radius of 160 km.

When small tubes are used in manometers, the capillary effects by either cohesion or adhesion of the fluids on the manometer walls need to be considered. The meniscus of a water-air-glass interface is concave upward while the meniscus of a mercury-air-glass interface is concave downward. Adhesive forces dominate in the former case while cohesive forces dominate in the latter case.

The rise in height readings of manometers due to capillarity is

$$h_c = \frac{2\sigma \cos \theta}{\rho g r}, \quad (8.11)$$

where θ is the angle between the manometer fluid and the glass, σ is the surface tension of the manometer fluid, and r is the radius of the manometer tube. When more than one fluid is involved, a correction factor is given by

$$C_c = \frac{2 \cos \theta}{\rho_m g} \left(\frac{\sigma[1 - m]}{r_1} - \frac{\sigma[2 - m]}{r_2} \right), \quad (8.12)$$

where m refers to the manometer fluid. Eq. (8.4) becomes

$$\Delta P = \rho_m g \left(\Delta h \left[1 + \frac{\rho_2 h_2}{\rho_m \Delta h} - \frac{\rho_1}{\rho_m} \left(\frac{h_1 + h_2}{\Delta h} + 1 \right) \right] \pm C_c \right). \quad (8.13)$$

Typical of values of surface tension and contact angles are given in Table 8.3.

For water-air-glass, the capillary correction is negative and for mercury-air-glass, the correction is positive. No correction is needed when the same fluid interface exists in both sides of the manometer or when the tubes are large (greater than 5 mm radius).

Table 8.3. Surface tension of water and mercury combination.

Fluid	Surface tension, σ (kg s^{-1})	Contact angle θ
Mercury-air-glass	0.470	140
Mercury-water-glass	0.380	140
Water-air-glass	0.073	0

8.2   Mercury Barometer

Evangelista Torricelli inverted a glass tube filled with mercury into a shallow dish of mercury in 1643. He noticed that the mercury in the tube remained about 760 mm above the mercury in the dish. This height varied from day to day and was attributed to the pressure exerted by the atmosphere on the earth.

Torricelli's apparatus was the forerunner of the modern day cistern-type barometer, the most common of which is the Fortin-type (Fig. 8.2). This barometer consists of a vacuum-referenced mercury column immersed in a large diameter cistern-type reservoir at atmospheric pressure. Thus, the dP in Eq. (8.2) is between atmospheric pressure and vacuum. The cistern is

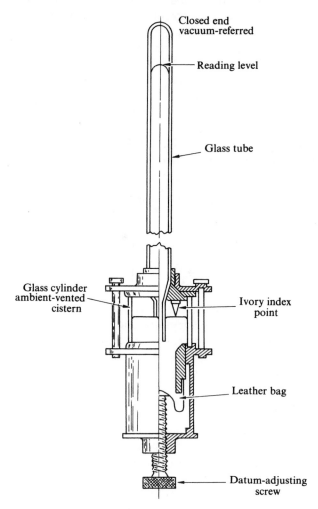

Figure 8.2. A Fortin-type barometer.

usually a leather bag in a housing with a level adjusting screw on the bottom and an ivory point to serve as the reference level (Anonymous, 1969; Middleton and Spilhaus, 1953).

In operation the level of mercury is raised to the reference level and the height of mercury in the tube is determined by observing the meniscus with a sliding vernier against a calibrated metal scale. The mercury column must be hanging vertically. In addition, the temperature of the barometer is required for mercury and scale corrections.

The atmospheric pressure observed in mm Hg is given by Eq. (8.2) as

$$P = \rho_{Hg} g h_{T_0}, \tag{8.14}$$

Table 8.4. Temperature corrections for Fortin barometer in mm Hg.

Temperature	Barometer reading (kPa)				
0°C	85	90	95	100	105
10	1.39	1.47	1.55	1.63	1.71
11	1.52	1.61	1.70	1.79	1.88
12	1.66	1.76	1.86	1.96	2.05
13	1.80	1.91	2.01	2.12	2.23
14	1.94	2.05	2.17	2.28	2.40
15	2.08	2.20	2.32	2.44	2.57
16	2.22	2.35	2.48	2.61	2.74
17	2.35	2.49	2.63	2.77	2.91
18	2.49	2.64	2.79	2.93	3.08
19	2.63	2.78	2.94	3.09	3.25
20	2.77	2.93	3.09	3.26	3.42
21	2.91	3.08	3.25	3.42	3.59
22	3.04	3.22	3.40	3.58	3.76
23	3.18	3.37	3.56	3.74	3.93
24	3.32	3.51	3.71	3.90	4.11
25	3.46	3.66	3.86	4.07	4.27
26	3.59	3.81	4.02	4.23	4.44
27	3.73	3.95	4.17	4.39	4.61
28	3.87	4.10	4.32	4.55	4.78
29	4.01	4.24	4.48	4.71	4.95
30	4.14	4.39	4.63	4.88	5.12
31	4.28	4.53	4.79	5.04	5.29
32	4.42	4.68	4.94	5.20	5.46
33	4.56	4.82	5.09	5.36	5.63
34	4.69	4.97	5.25	5.52	5.80
35	4.83	5.11	5.40	5.68	5.97

where h_{T_0} is the height of mercury at a standard temperature and elevation. The height of mercury is corrected for the linear coefficient of thermal expansion of the brass scale, S, and the cubical expansion of mercury, m, by

$$C_T = \left(\frac{S(T - T_s) - m(T - T_0)}{1 + m(T - T_0)}\right)h_T, \qquad (8.15)$$

where T_s is the temperature at which the scale was calibrated, T_0 is the reference temperature, and T is the indicated temperature. The scales of Fortin and other barometers commonly are made of yellow brass, the linear expansion coefficient of which is $18.4 \times 10^{-6} \, {}^\circ C^{-1}$. Using $181.8 \times 10^{-6} \, {}^\circ C^{-1}$ for the cubical expansion of mercury, Eq. (8.15) becomes

$$C_T = -\frac{163.4 \times 10^{-6} T h_T}{1 + 181.8 \times 10^{-6} T}. \qquad (8.16)$$

Selected values of C_T for brass scales calibrated at $16.7°C$ with reference temperature of $0°C$ as given by Brombacker et al. (1960) are shown in Table 8.4.

In addition to the temperature correction, the barometer reading must be corrected for gravity and elevation by using Eqs. (8.9) and (8.10). Equation (8.14) becomes

$$P = \rho_{Hg}[g_\phi - 3.086 \times 10^{-6}h + 1.118 \times 10^{-6}(h - h')](h_T + C_T). \qquad (8.17)$$

8.3   Aneroid Barometer

Though mercury barometers are very accurate, they are not easily automated. The desire for an electrical signal which can be recorded leads us to the aneroid barometer.

In early multiple aneroid barometers, each bellows contained an inner spring (Fig. 8.3.a) and had to be individually evacuated. The partially evacuated bellows would vary in thickness with changes in atmospheric pressure. The inner springs caused an uncertain amount of friction. The performance of barometers was improved by constructing the bellows from tempered steel which acts also as the spring (Fig. 8.3.b). The entire system could be evacuated through a single tube. Most modern aneroid barometers contain an assembly of flexible metal bellows constructed from beryllium copper or Ni Span-C alloy. The more sensitive barometers contain as many as 14 bellows.

Electrical signals proportional to pressure are obtained with the use of potentiometers or linear variable differential transformers. Resolution of 10 Pa and linearity of 50 Pa is obtainable over a narrow range (for example, 10 k Pa). The temperature coefficient is of the order of $0.005\% \, {}^\circ C^{-1}$.

The use of potentiometers as voltage dividers is described in Sect. 2.2. However, the linear variable differential transformer requires additional

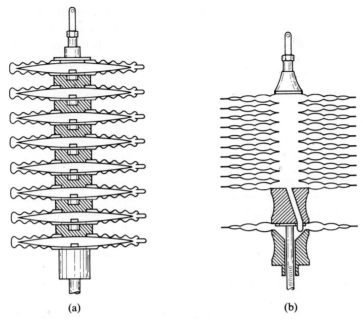

(a)   (b)

Figure 8.3. Multiple aneroid cells for a barograph: (a) constructed with internal springs; (b) bellows constructed from tempered steel that also acts as the spring.

explanation. The schematic diagram of the transducer is shown in Fig. 8.4. It consists of three coils, P, S_1, and S_2, and a core of magnetic material. In use, 6.3 V at 60 Hz is applied to the coil P, and the current induces voltages E_1 and E_2 in the identical coils S_1 and S_2. At symmetrical balance the output signal, $E_o = E_1 - E_2$, is zero. Displacement of the core causes magnetic asymmetry and an output signal. The output increases linearly over a considerable range as the core is moved and changes polarity as the core is moved through the zero position (Lion, 1959).

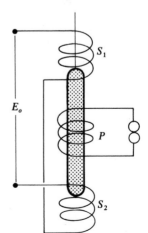

Figure 8.4. Schematic of a linear variable differential transformer in which S_1, S_2, and P are coils, C is a core of magnetic material, and E_0 is the voltage out.

The sensitivity of commercial transducers varies from 0.2 to 1.2 mV for a displacement of 0.01 mm V^{-1} applied to the primary coil. The variable differential transformer is rugged and mechanically simple, suitable for displacements from 0.002 mm to several centimeters with 0.1 to 0.3 g of applied force. The output is relatively high and the impedance is low to medium. This transducer makes it easy to obtain an electrical signal from the aneroid barometer.

Literature Cited

Anonymous (1969). Handbook of Meteorological Instruments. Part 1. Instruments for Surface Observations. Her Majesty's Stationery Office. London, 458 pp.

Benedict, R. P. (1977). Fundamentals of Temperature, Pressure, and Flow Measurements. Wiley, New York. 517 pp.

Brombacher, W. G., D. P. Johnson, and T. L. Cross (1960). Mercury barometers and manometers. Nat. Bur. of Stand. U.S. Monogr. **8**.

Lion, K. S. (1959). Instrumentation in Scientific Research. p. 48. McGraw-Hill, New York. 323 pp.

Middleton, W. E. K. and A. F. Spilhaus (1953). Meteorological Instruments. Univ. Toronto Press, Toronto, Canada. 286 pp.

Chapter 9

Data Acquisition Concepts

The various physical and electrical quantities discussed in this book contain information that the different measurement transducers and techniques seek to define. The physical quantities (temperature, heat flux density, wind velocity, pressure, etc.) have generally been converted into electrical signals which, in turn, are recorded and interpreted. The conversion has been accomplished by a variety of transducers. The information that is presented to the transducer and passed on by it (in converted form) can be classified as to whether the information is in electrical or nonelectrical form.

Information that is transformed into electrical quantities may be represented by several characteristics. These characteristics include the amplitude of the electrical quantity (voltage, current, resistance, etc.), the frequency, the duration, and the presence or absence of the quantity. Malmstadt et al. (1974) grouped the information associated with the various quantities into four *data domains*. The *physical domain* (P) includes all nonelectrical signals, for example, temperature. The *analog domain* (A) includes the information associated with amplitude of electrical signals that are continuous functions of time. The *time interval domain* (Δt) includes the information associated with frequency or pulse width. The *digital domain* (D) includes information that is a discontinuous function of time, but that is related in integer steps to the desired quantity. The transducers serve to convert the information from one domain to another. All combinations of conversions between P–A–Δt–D are possible. However, we most commonly utilize conversions P–A (example, thermocouples) or P–Δt (e.g., pulse generation by anemometers or tipping-bucket rain gauges).

9.1   Signal Characteristics

The distinguishing feature of analog signals from the analog domain is their continuous nature. Analog signals may be periodic, recurring at specific intervals, or nonperiodic, depending on the nature of the phenomena that generates them. Analog signals may be recorded continuously (various chart recorders) at appropriate degrees of resolution, or they may be sampled periodically. With proper sampling, the original analog signal can be reconstructed. The basic sampling theorem is discussed in most electrical engineering texts; we shall consider in a subsequent section some selected aspects of sampling variable signals.

Digital signals, in contrast to analog signals, are discontinuous and are characterized by presence or absence (on–off). The presence or absence of pulses in a pulse train may contain the information that is desired. A common form of a pulse train in communication equipment is the *pulse-amplitude modulated* (PAM) signal, in which the amplitude of the pulses varies with time. The function that modulates the amplitude may be a sine wave, for example, and information is carried on both the modulating signal and the pulse train.

The time domain signals include frequencies of the continuous analog signals, the rate at which pulses in a pulse train arrive at a specific point, or the width of the pulses.

Until recently, signals that were transformed from the physical domain into the analog domain were recorded on analog recorders and eventually transformed into usable digital form by visual inspection and interpretation (the infamous "optical encoder"). Now, however, various techniques are available to transform the analog signals into digital form in one continuous operation, using rather complex systems for converting, transferring, and handling the desired information.

9.1.1   Representation of Digital Data

One simple way to represent a digital signal is by the use of 10 separate data lines, each representing an integer from 0 through 9. If each line is connected through a switch to a power source, and is further equipped with some means for determining whether the switch is open or closed, then the closure of one switch is all that is necessary to indicate any digit between 0 and 9. One simple way to sense the potential on each line is by adding a small light bulb in series with the switch. The light will indicate to the observer which digit is selected in this decimal representation. The switches, of course, might be closed mechanically in the simplest demonstration of the decimal representation, or they might be opened or closed electronically.

Note, however, that nine lines are unused whenever a single digit is being presented. This is an inefficient way to represent digital data. Only four lines are needed to represent a digit in the *binary* system representing a considerable reduction in wiring and in hardware (wire, switches, lights or other indicators, etc.). In the decimal system, each line had a "weight" of one; in the binary system, each line has a different weight (commonly 1, 2, 4, or 8), and the lines are combined as necessary to represent an integer. For example, if the lines with weights 1, 2, and 4 are "on" the number 7 is indicated. If all four lines are on, the weights used in this example would total 15. If instead of 4 lines, we had 10, we could represent any number between 0 and 512 in binary representation. If we restrict our lines to 4, and limit the combinations that can be energized, we can present any number from 0 to 9 in widely used *binary-coded-decimal* (BCD) which combines some of the desirable data-handling characteristics of the binary system with the familiar and readily interpreted characteristics of the decimal system.

The binary representation also provides a simple way to minimize data errors by adding a fifth line designated as the parity-check line. Note that an even number of lines are on whenever a 3, 5, 6, or 9 is represented in binary form, while an odd number of lines represents a 1, 2, 4, 7, or 8. If we turn the parity-check line on whenever a 1, 2, 4, 7, or 8 is being presented and off for the others, the parity signal makes it easy to see if one of the lines has been inadvertently turned off (or turned on by mistake). Even parity sets the parity check on when there is an even total of lines, e.g., the numbers 3, 5, 6, or 9, while odd parity is the reverse. We are assuming that the probability of two lines being in error is very, very small. A parity check with a single line could not detect the presence of two simultaneous errors.

These examples have considered only data presented simultaneously in parallel to four (or ten) separate points. The data could easily be presented sequentially in *serial* form as a succession of pulses arriving at one point. The presence, or absence, of pulses or *bits* can be interpreted in the same way as signals presented over data lines in parallel. The absence of pulses is determined by comparison against clock time.

Only four bits are needed to transmit a single digit. However, both the sender and the receiver will have to remain in near-perfect synchronization in order for the signals to be interpreted properly. This *synchronous* transmission can be accomplished if provision is made for a separate clock signal to keep both units on time. The same effect can be achieved in *asynchronous* transmission by simply bracketing the four data bits with "start" and "stop" bits so that the two communicating devices need remain in synchronization for only a short time. The start bit warns the receiver to prepare for data and gives it a reference for clock synchronization. The one or two stop bits following the data allow the receiver time to recover before the next character arrives.

The data transfer rate, or *Baud rate*, describes the number of data bits transferred serially per second. We noted earlier that only four bits are needed to transfer an integer, plus additional bits for parity, for starting, and for stopping. The combination of data bits is defined in the American Standards

Association's code for communications, known as ASCII, where seven bits define a character. The 127 recognized combinations of the seven data bits may represent an integer, an upper or lower case alphabetic character, a special symbol, or a control message character that is not usually displayed by the receiver. Once the total number of bits is known, the number of characters transmitted per second can be calculated from the Baud rate. Characters are frequently represented by 11 bits (7 data bits, 1 parity, 1 start, and 2 stop bits). The common teletype is rated at 110 Baud; this is the equivalent to 10 11-bit characters per second. Standard data systems now transfer data at rates that range up to 9 600 Baud or more.

9.1.2   Analog-to-digital Conversion

We shall briefly examine three of the many methods that have proven useful in converting analog data into digital form. One of the methods involves a direct conversion from the analog domain into the digital domain. The other two methods involve first a conversion into the time interval domain, and then a conversion into the digital domain. All three methods are widely used in commercial instruments. Although very sophisticated techniques are used in the actual instruments, the principles discussed here illustrate their operation.

The successive approximation technique is the first conversion scheme; it has proven quite successful. This method matches the unknown voltage against a known voltage that is obtained from a digital-to-analog (D–A) converter. The D–A converter may be a simple resistor network, in which precision resistors are switched in or out of a divider circuit in accord with the digital input settings. The value of the unknown voltage is first "fixed" with a "sample and hold" circuit. The successive approximation comparisons then begin with the most significant digit in the D–A converter being varied bit by bit. The value of the most significant digit is saved at the point that the output of the D–A converter first falls below the value of the unknown voltage. The next most significant digit is then tested and its value saved at the point that the D–A output first falls below the unknown voltage. The process continues until the value of the least significant digit is established, and at that time the value of the unknown voltage is given by the setting on the D–A converter. The resolution of the reading depends on the resolution of the D–A converter. The system is fast (in the range 1–10 ms per reading), relatively inexpensive, and quite common. The initial instantaneous reading of the unknown voltage is subject to noise (see Sect. 9.4) unless the signal is filtered. The filtering will negate the speed of the successive approximation method to some extent.

The other two schemes for converting analog data to digital form pass through the time interval domain. These methods are based on the proportionality that exists between the difference in two voltage levels and the time required for a "linear ramp" signal to change between them. The time interval (Δt) is readily transformed into a digital word by transferring clock pulses at a constant rate into a counter for the duration of the interval. The ramp signal can be used directly, as in the single-slope linear ramp method, or it can be used indirectly, as by the voltage-controlled oscillator.

The single-slope linear ramp measures the time required (Δt) for the ramp to rise from zero to the value of the unknown voltage. The system requires a stable, fixed ramp generator, a sensitive comparator to test for equality of the voltages, and a pulse generator and counter. The single-slope technique is not used extensively because it has rather low accuracy (typically 0.1–1%) and is somewhat sensitive to noise on the input signal. These deficiencies are minimized by the dual-slope integration modification. The unknown voltage is first integrated for a fixed time interval Δt_1, using a standard integration circuit that charges a capacitor at a rate proportional to the level of the unknown voltage. At the end of period Δt_1, the partially charged capacitor is switched to a reference voltage of negative polarity and allowed to discharge to zero over period Δt_2. The count of clock pulses during the period Δt_2 is proportional to the unknown voltage. The noise rejection characteristics of this method are excellent because the input voltage is integrated over the fixed period Δt_1. The accuracies approach 0.05%, with a period of about 100 ms required for completing the digitizing cycle.

The third technique, also based on the A–Δt–D conversion, uses a voltage-controlled oscillator (VCO) to produce a frequency that is proportional to the unknown input voltage. The unknown voltage is input to a conventional integration circuit in the VCO, as in the dual-slope integration method, but the capacitor is charged only to a reference voltage level before it is discharged. The rate at which the capacitor is charged and discharged is proportional to the unknown voltage. The charge–discharge frequency is input directly to a counter. The longer the period for which the frequency is totalized (integrated), the greater the noise rejection capabilities. Accuracies with this technique are high, and can approach 0.005% for integration periods in the order of 167 ms.

9.2   Digital Data Acquisition Systems

The precision and speed of the analog-to-digital converter is brought to bear on field measurements with the addition of components that sample the inputs, transfer data, and record the digital data. Standard, off-the-shelf data acquisition systems are available with high precision, small size, and excellent reliability. Those now commonly used for environmental measurements have resolutions in the order of 1 part in 1 000 000 and with excellent

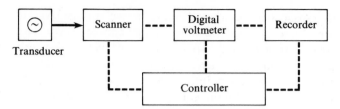

Figure 9.1. Idealized digital system.

accuracy. Most environmental systems are of moderate speed and scan among the inputs at rates of 1 to 40 channels per second. Long-term reliability has proven excellent on many of the systems now in use. Size and power consumption are small, and it is relatively easy to expand most of the available systems to hundreds of channels.

The basic components in a digital system are shown in Fig. 9.1. The components form a chain, transducer—scanner—D–A converter—recorder, all associated with a system controller. The electrical signals from the transducers are presented to a scanning unit. The scanner selects the appropriate inputs and transfers the signal from a given instrument to the A–D converter. The digitized data is transferred to the digital recorder for eventual processing and analysis. The components are linked to a controller that synchronizes the various operations and data transfers. The controller normally includes a clock; it may now also include either a minicomputer, a microprocessor, or a calculator for on-line processing and analysis.

The trend in the past decade has been toward systems with lower power drains, higher speeds, increased accuracy, and greater reliability. The slower recorders (such as teleprinters, paper tape punches, etc.) have been superseded by several types of magnetic tape cassette units, supplemented with high-speed printers. The increasing capability and availability of microprocessors will bring on-line data analysis into more and more field experiments.

The rapid development in digital components has generated a great deal of interest in standardization. This has led to recent development and general acceptance of a standardized interface to link the various components in a digital system together under the direction of the controller. This interface is defined in IEEE Standard 488, and it is referred to as the General Purpose Interface Bus (GPIB). The GPIB definition classifies system elements as *controllers*, *talkers*, or *listeners*. The controller has the power to designate which peripheral unit will talk (only one talker at a time) and which will listen (up to 14 at a time). The data transmission format, timing, and protocol for sending and receiving digital messages are all defined for GPIB. The IEEE Standard 488 interface definition specifies those mechanical, electrical, and functional elements that are independent of the devices to be connected. The device-dependent operational elements are not specified, leaving a great deal of flexibility for system design.

The general adoption of this standard has given an additional impetus to the incorporation of controllers into systems that have previously been made up solely of talkers and listeners.

9.3   Some Sampling Considerations

It is difficult to obtain signals that have not been modified in some way during the transformation from physical to electrical form. There is invariably some change introduced by the transducer; the reduction in signal amplitude and phase was defined earlier, in Sect. 3.2, with the concept of the *time constant*. These changes apply to all signals, regardless of whether they are being recorded continuously (as with a single-channel recording potentiometer) or are sampled periodically (as with a multichannel potentiometer or a digital data system). The recording system may introduce an additional modification of the original signal. For analog recorders, the speed of response (the recorder's time constant) may further reduce the amplitude and delay the phase of the signal from the transducer. The possible changes introduced by recorders that sample at discrete intervals are somewhat more subtle.

We shall look in more detail at the problem of sampling a variable signal at discrete time intervals (Δt, s). The problem can be stated quite simply: Which Δt sampling interval should be selected in order to fully define a given input signal? The requirements of "fully defining" a signal will vary in practice, e.g., examination of mean amplitude will require considerably less frequent sampling than will an analysis of the frequency of signal fluctuations. At this point, however, we will examine the sampling required to fully define, i.e., reconstruct, the signal

The theory of sampling periodic signals has been well worked out in electrical engineering; most textbooks in this field discuss *Shannon's sampling theorem*. The theorem can be restated here for our purpose as:

> Let $p(t)$, a signal giving the true relation between p and t, have a maximum frequency component of v_m. Let values of $p(t)$ be determined at regular intervals separated by time $\Delta t_s \leq 1/2v_m$. The signal is periodically sampled every Δt_s. The samples $p(n\,\Delta t_s)$, where n is an integer, uniquely determine the signal and the signal can be reconstructed from these samples without error.

The time Δt_s is the *sampling time*, and the *sampling rate* is defined by $v_s = 1/\Delta t_s$. Since the theorem states that $v_s \geq 2v_m$, at least two samples are taken in the course of the period corresponding to the highest frequency component of $p(t)$. The minimum sampling frequency to permit reconstruction of the signal, $v_s = 2v_m$, is called the *Nyquist frequency*.

When the signal is sampled at a frequency less than the Nyquist frequency, *aliasing* will occur. Aliasing is the addition of lower frequency signal components to the original when the reconstruction is carried out. If, for example, a signal is sampled at too slow a rate (say $v' \leq v_s$), the signal that is reconstructed will have lower frequency components ($\frac{1}{2}v'$) than that of the original ($\frac{1}{2}v_s$). While it is of critical importance in communications to precisely reconstruct the original signal, it is seldom of importance in environmental measurements. The major consequences of aliasing errors will occur only when analyses are being made of the frequencies of the reconstructed signals, as in spectral analyses.

The sampling theorem has been used, however, as the basis for estimating the frequency with which a signal should be sampled. These considerations have sought to simplify the problem of specifying the sampling frequency, which requires knowledge of the highest frequency components in the physical signals, by examining the time constant of the transducer that generates the electrical signals. This consideration recognizes that the transducer filters out the highest frequency signals and that the problem is one of sampling the output of the transducer rather than the physical input to the transducer.

In perhaps the first discussion of optimum sampling frequencies for digital sampling systems used with environmental signals, Tanner (1963) concluded that Shannon's theorem justified the rule of thumb that the time interval between successive samples should be equal or less than twice the time constant of the transducer ($\Delta t = 2\tau$). Fuchs (1971), in reviewing several proposed criteria for selecting the sampling interval, concluded that it should be $\alpha\pi\tau$, with α selected to be a desirably small attenuation factor. The factor is defined earlier, in Eq. 3.7i, as the attenuation expected in a periodic signal that is "filtered" by an instrument with time constant τ. If $\alpha = (1 + \omega^2\tau^2)^{-1/2}$ is solved for a "cutoff" frequency v_c (noting that $\omega = 2\pi v$), for any given value of α, we can define the frequency below which signals will be rejected by a sensor of time constant τ, with respect to an arbitrarily selected criteria for attenuation. Fuchs noted that there were no good rules for selecting an appropriate value of α to be used in solving for v_c. This selection would, in effect, specify the magnitude of the maximum frequency component v_m in the sampling theorem and thus requires that prior knowledge be available on the characteristics of the signals to be sampled.

What are the consequence of sampling sensors with time constants τ, at periodic intervals Δt? If we follow Fuchs' (1971) recommendation that $\Delta t = \alpha\pi\tau$, and if we arbitrarily select a value of $\alpha = 0.05$ as the level of short period signals that are acceptable, we can calculate the critical sampling frequency v_c that will ensure that no information is lost. From the definition of α, we calculate $v_c = 3.18\tau^{-1}$. The sampling frequency is then $v_s = 2v_c = 6.36\tau^{-1}$, which corresponds to a period between samples of $\Delta t = 0.157\tau$. In other words, with this value of α, we must sample at intervals of 0.157τ to obtain all of the available information from periods greater than 0.318τ in the sensor output. Using this criteria, and accepting all signals transmitted by the transducer with an amplitude of 0.05 or greater, it will be necessary to sample at a time interval slightly smaller than 1/6 of the transducer time constant.

9.4   Signals and Noise

The output signal from each transducer is defined by the type of transducer and the relationship governing its transformation of physical signals into electrical signals. Distortion of the true output signal can arise from a variety of external influences; we will refer to this distortion as "noise." The distortion may in some instances be solely in amplitude, or it may also include distortion of the wave shape or of the frequency in rapidly varying signals.

The distortion may be associated with many causes, from those within the transducer itself to those within the final transformation and recording of the output signal. The sources of distortion will generally fall within one of two groups: (1) noise from external sources that enters the circuit through electromagnetic radiation or through some direct connection and (2) noise that may be generated by elements within the circuitry of the recording system.

We shall examine noise from external sources in some detail because careful attention to a few simple guidelines can markedly improve the quality of data obtained. Noise generated by elements within the recording system can be equally troublesome, but it may not be easy to isolate and correct, particularly when it results from a design weakness in a commercial recorder.

Let us first define normal mode and common mode signals. The distinction is simple. A *normal mode* signal appears at only one of the two terminals that are output from the transducer (or are input to a recorder). A *common mode signal*, in contrast, appears simultaneously at both terminals.

A frequently encountered measurement problem is that of measuring a small normal mode signal in the presence of a large common mode signal. It might, for example, involve measurement of the voltage drop across a resistor in the high voltage side of an electrical circuit. If the potential on the high side is only a few volts and if the recorder is well-designed, there is little problem in precisely measuring even rather small normal mode signals. Each recorder design, though, responds differently to various common mode signals. Problems arise if the recorder allows leakage from one terminal to ground, for then the current to ground generates an offset voltage that is applied to one terminal and appears as noise on the normal mode signal. The leakage problem increases as the common mode signal level increases. It also increases as the common mode signal changes from dc to ac because at higher frequencies current will be transferred from the terminals to ground by capacitive coupling. If the two terminals provide a balanced pathway to ground, the noise applied to each terminal will cancel leaving the correct normal mode signal. The extent of the conversion of common mode signals to normal mode noise thus becomes a question of design.

Most recorders are designed and physically constructed to provide a balanced leakage path from the terminals to ground to minimize errors arising from common mode signals. It is almost impossible to obtain identical leakage paths over all frequencies, however, and some of the common mode signal appears as noise. The instrument's performance in making the correct normal mode readings in the presence of a common mode signal is expressed by the *common mode rejection* (CMR), commonly defined in *decibels* (dB). CMR is a ratio, calculated in decibels as

$$\text{CMR(dB)} = 20 \log_{10} \frac{\text{common mode signal}}{\text{normal mode noise}}. \tag{9.1}$$

A CMR of 120 dB indicates that the common mode signal is 10^6 greater than that of the common mode noise that appears. A recorder with these specifications would show 1 μV error for 1 V of applied common mode signal.

The rejection ratios can be quite large. A good quality digital system may have a CMR of 160 dB or greater, meaning that it will pick up 1 μV error from a 100 V common mode signal. The manufacturer's specifications should indicate the frequencies over which the rejection ratio applies. In practice, the specifications for a given recorder are frequently given in terms of the overall effective common mode rejection, a figure that combines CMR with a similar normal mode rejection.

The normal mode rejection, calculated in a manner similar to Eq. (9.1), can be increased by design of the digital voltmeter (normal mode rejection is enhanced by the integration feature of the dual slope integrator and the voltage–frequency designs discussed in Sect. 9.1.2) and by use of a low-pass filter at the input of the recorder.

As an example, the effective overall common mode rejection of a commercial digital voltmeter[1] of voltage–frequency design is, with up to 250 V common mode voltage and 1 000 Ω source unbalance, greater than 180 dB with dc and at power frequencies (50 or 60 Hz, $\pm 0.1\%$), and greater than 140 dB from dc to 2 kHz. Note that the specification includes the frequency ranges that are applicable and includes the amount of unbalanced resistance that can be in the circuit. The consequences of circuit imbalance are discussed further in Sect. 9.4.2.

9.4.1   Electrical Interference

The signal leads that link the transducer to the recorder often serve as a pathway through which electrical interference can become incorporated into the desired signal. The basic mechanisms involved include electromagnetic induction and electrostatic coupling. Transfer of electrical energy by either can occur as the result of the presence of the signal leads in an electric field or as the result of current through or over the signal leads through a direct connection (a ground loop). These problems are illustrated in Fig. 9.2. We will discuss first the problems associated with the signal leads serving as a conductor in an electric field and then examine the noise problems associated with ground loops.

---

[1] Specifications. Autodata Nine with A904 high resolution digital voltmeter. Acurex Autodata, 485 Clyde Avenue, Mountain View, California 94 042.

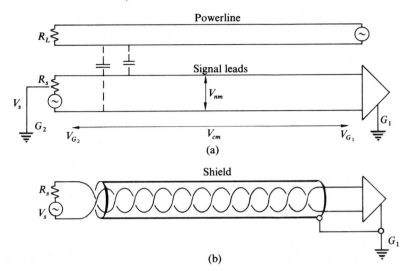

Figure 9.2(a) and **9.2(b)**. Noise pickup and rejection. (a) Noise pickup will occur by induction, capacitive coupling, and possible group loop currents. Symbols are defined as R_L, load resistance; R_s, source resistance; V_s, source voltage; V_{nm}, normal mode voltage; V_{cm}, common mode voltage; V_G, ground potential; and G, ground. (b) The twisted, shielded pair rejects inductive and capacitive pickup. The input is floating and there is only one path to ground.

Any current-carrying wire is surrounded by electromagnetic lines of induction that can generate an ac current in an adjacent signal cable. The voltage associated with the induced current is a function of the area of the loop made by the signal cable as it runs to the transducer and back to the recorder, the distance between the signal cable and the power cable, and the rate at which the magnetic field changes, i.e., the frequency of the alternating current in the power cable. The induced current generates a normal mode ac voltage in the signal cable that is applied to the recorder terminal in the same manner as the desired signal.

Electromagnetic noise transfer between the signal leads and a power cable can be minimized at most environmental measurement sites. First, the signal cables should always consist of "twisted-pair" configuration; twisted leads minimize the loop area for pickup, and the inductive pickup tends to be self-cancelling. Second, the intensity of the magnetic field is inversely proportional to the cube of the distance between the power cable and the signal leads, so inductive pickup is minimized by physically separating the power cable and signal leads. Any needed cable crossings should be made at right angles. Third, the induced ac voltage is of normal mode, and it can be filtered if needed, using a low-pass filter that rejects the ac component and passes the desired dc component. Finally, special metallic shielding of high magnetic permeability (ferromagnetic material) could be used to enclose the signal leads and provide a good magnetic path for shunting the lines of induction away from the input loop. While this special shielding has proven useful in some industrial environments, it is unlikely that such extreme measures will be needed for environmental measurements.

The signal leads will also be electrostatically coupled to other circuits at different potentials. The transfer of current to the signal leads depends on the difference in potential and on the capacitance that exists between the leads and other circuits. The capacitance is a function of the effective dielectric separating the leads and the other circuits, the distance between them, and the physical dimensions of the affected portions of the measuring circuit.

The pickup of radio frequency waves can be troublesome. The signal leads serve as an antenna whose efficiency depends on the frequency of the waves, the size and length of the leads, and the impedance to ground. Electrostatic pickup is coupled to each side of the signal leads and presented to the recorder as a common mode signal.

Electrostatic pickup can also originate from charged surfaces and objects near the site of the measurements. Various objects can develop substantial static charges, particularly in dry conditions. Disturbances can result from the movement of a charged object into the area, such as when the person making the measurement is wearing clothes charged with static electricity. Also, motion of hands and arms can affect the normally constant values of capacitance that exist between the instruments and the environment, causing changes in the electrostatic transfer and generating erratic excursions in the recorded signal.

Electrostatic pickup is most easily eliminated by enclosing the signal leads within a shield. Most modern amplifiers are already enclosed within a guard shield. In recent years, the braided metallic shields for signal leads have been almost completely replaced with a metallic foil shield that provides a high degree of protection. The braided shields are more flexible, however. The foil shields are equipped with a ground wire. In practice, the shield and the amplifier guard are both grounded at the same point and provide a shunt to transfer the electrostatic energy harmlessly away from the recording circuitry. The electrostatic common mode signal present on a pair of unshielded or poorly shielded leads can be substantially reduced by coupling the two signal leads together with an appropriate capacitor.

Ground loops can be another source of noise. They cause problems by permitting unwanted currents to flow between separate points in the circuit. Consider, for example, measurements made with a grounded transducer when the recorder is grounded separately, at yet a different point. Even if the two ground points are at the same potential, a loop exists for the pickup of inductive signals, with the pair of signal leads providing one side of the loop and the return path between the two grounding points providing the other. The induced current will be presented as a common mode signal to the two recorder inputs; the twisted signal pair will not minimize this type of induction as both leads serve as one side of the input loop.

Several other ground loop configurations will generate the same type of common mode signal. It is very likely that the ground potentials at the two points in the previous example will be at different, rather than equal, potentials. In this case, current will flow down the signal leads and appear as a common mode signal at the recorder, simply as a consequence of the differing potentials at the two ground points.

Another ground loop problem occurs when the transducer is floating, but either the shield is grounded at both ends, or at the transducer end and at the recorder. As a consequence of differing potentials, current will flow down the shield and can transfer current through the capacitive coupling that exists between the charged shield and the signal leads.

The remedy for ground loop problems is simple. Break all ground connections except one. Typically, the transducer in environmental measurements will be floating instead of grounded (thermocouple in air, for example), and the ground point for the shield and recorder will be at the recorder end. If there is leakage between the transducer and ground (as in a soil thermocouple with defective insulation), the possibility exists for ground loop problems. In some cases, it is desirable to ground the transducer and allow the recorder to float.

9.4.2   Other Sources of Noise

There are a number of additional sources of noise, most of which are of minor consequence. However, appreciable error can develop from high source resistances and from thermal offsets generated within various circuit components.

The "circuit loading error" associated with a high source resistance was discussed earlier, in Chapter 2. This error becomes significant when the source resistance rises to an appreciable fraction, say 1 %, of the input resistance of the recorder. It is more common with older recorders whose input resistances are generally lower than those of modern instruments.

High source resistance can contribute other detrimental effects as well. In configurations that permit common mode currents to circulate through the signal cable loop, the current flowing through the source resistance is converted to a normal mode signal and appears as a voltage offset to the recorder. For this reason, radio frequency pickup increases as the source resistance increases. The results can be troublesome, even when the currents are small. For example, a 10 μV offset will be generated by a current of only 1 nA flowing through a source resistance of 10 kΩ.

Additional voltage offsets will be generated by common mode currents flowing through resistances in each side of the signal cable. If the two sides are balanced with respect to resistance, the offsets will cancel. It is more likely, however, that the two leads will be unbalanced, either in the line or at the source, and the net effect will be another voltage offset. For these reasons, it is prudent to keep the source resistance as small as is feasible, and to minimize any resistance inbalance in the signal leads.

The thermoelectric effects discussed in Chapter 3 will occur at the junction of any two dissimilar metals. Thus, thermal offset voltages can develop in measuring circuits and disturb precision measurements of low-level signals. The presence of thermal offsets can be detected easily, either by heating the various circuit components, or by cooling them with certain volatile substances that are packaged for this purpose in aerosol spray cans.

Thermal offsets can be minimized by using copper-to-copper or noble metal contacts. The copper surfaces must be bright, however, for copper versus copper oxide has appreciable thermoelectric power. Also, temperature changes in the circuit elements can be minimized by insulation, thermal lagging, or maintenance of a constant-temperature environment.

9.4.3   Minimizing the Effects of Noise

Two simple checks are sufficient to determine the presence of most types of noise problems: first transmit a zero signal and then transmit a known signal while observing the recorder output. If unexpected results are observed, the cause may well be noise of one form or another. The zero or null signal can be obtained for many transducers by shorting the output terminals; the known signal could be generated in a temperature transducer, for example, by immersing the element in an ice bath or constant temperature medium. The results observed should fall within the calibration tolerance on the measuring system. It is advantageous to generate the known signal with the transducer that is to be used in the actual measurement program in order to pick up any problems that may be specific to that particular transducer. Even if no problems are evident, it is good practice to devote one extra measurement channel to a null obtained by shorting the ends of an unused signal pair and one other unused channel to a known signal, such as from a thermocouple in an ice bath. The two additional measurements will indicate the overall functioning of the measurement system and will help to isolate any problems that develop during an observation program.

As another check on the overall functioning of the system, one should grasp the signal cables near the recorder while observing the displayed signal. An excursion will indicate possible problems with grounding or shielding. If problems exist, do not overlook the removable grounding link used on many recorders to connect the guard on the amplifier to the ground lug. Each manufacturer has specific instruction on the conditions that require this link to be either connected or open. If these are not available, try out both ways. The actual ground must be of good quality and made only at a single point, usually at the recorder end. It is sometimes advantageous to bypass the normal ground point in the standard 3-wire power cord, and provide a substitute connection with heavy copper wire tied directly to a cold water pipe. In the field, a suitable ground can usually be obtained by driving a standard ground rod (copper-plated steel) into moist earth. In a dry area, it may be necessary to saturate the ground around the driven rod to establish a satisfactory ground.

All signal cables should be twisted-pair, well-insulated, and shielded against electrostatic pickup. The shields should be connected in accordance with the recorder manufacturer's instructions. Generally, it is better in digital systems to connect the shields through the scanner so that they are switched along with the signal pair. The manufacturer will provide a ground path through the scanner for this configuration. The shields should otherwise be tied to a common ground point. Some systems require that the transducer end of the shield be tied to the low lead of the signal pair while other configurations will leave that end floating. Lapped-foil shields with a copper drain wire give the best coverage and protection against electrostatic pickup. Exposed portions of the signal cables should be kept as short as possible, not more than a centimeter or so beyond the shield. The shields should be carried through connectors in the signal leads. The shield lug on some connectors is tied to the case and use of this lug can result in an inadvertent multiple ground if the connector is allowed to contact the instrument support or tower.

Separate the signal cables from the power cables to minimize capacitive and inductive pickup, and arrange for the cables to cross at right angles if a crossing is necessary. Careful attention to these simple wiring, grounding, and shielding notes will minimize many otherwise troublesome noise problems.

Bibliography

Fuchs, M. (1971). Data logging and scanning rate considerations in micrometeorological experiments–a discussion. *Agric. Meteorol.* **9**:285–286.

Malmstadt, H. V., C. G. Enke, and S. R. Crouch (1974). Electronic Measurements for Scientists. Benjamin, New York. 906 pp.

Tanner, C. B. (1963). Basic Instrumentation and Measurements for Plant Environment and Micrometeorology. Dept. Soils Bull. 6, Univ. Wisconsin, Madison, Wisc.

Index